EQUATIONS, INEQUALITIES, & VIC's

Math Preparation Guide

This essential guide covers algebra in all its various forms (and disguises) on the GMAT. Master fundamental techniques and nuanced strategies to help you solve for unknown variables of every type.

Equations, Inequalities, and VIC's GMAT Preparation Guide, 2007 Edition

10-digit International Standard Book Number: 0-9790175-2-1
13-digit International Standard Book Number: 978-0-9790175-2-0

8 GUIDE INSTRUCTIONAL SERIES

Math GMAT Preparation Guides

Number Properties (ISBN: 978-0-9790175-0-6)

Fractions, Decimals, & Percents (ISBN: 978-0-9790175-1-3)

Equations, Inequalities, & VIC's (ISBN: 978-0-9790175-2-0)

Word Translations (ISBN: 978-0-9790175-3-7)

Geometry (ISBN: 978-0-9790175-4-4)

Verbal GMAT Preparation Guides

Critical Reasoning (ISBN: 978-0-9790175-5-1)

Reading Comprehension (ISBN: 978-0-9790175-6-8)

Sentence Correction (ISBN: 978-0-9790175-7-5)

HOW OUR GMAT PREP BOOKS ARE DIFFERENT

One of our core beliefs at Manhattan GMAT is that a curriculum should be more than just a guidebook of tricks and tips. Scoring well on the GMAT requires a curriculum that builds true content knowledge and understanding. Skim through this guide and this is what you will see:

You will *not* find page after page of guessing techniques.

Instead, you will find a highly organized and structured guide that actually teaches you the content you need to know to do well on the GMAT.

You *will* find many more pages-per-topic than in all-in-one tomes.

Each chapter covers one specific topic area in-depth, explaining key concepts, detailing in-depth strategies, and building specific skills through Manhattan GMAT's *In-Action* problem sets (with comprehensive explanations). Why are there 8 guides? Each of the 8 books (5 Math, 3 Verbal) covers a major content area in extensive depth, allowing you to delve into each topic in great detail. In addition, you may purchase only those guides that pertain to those areas in which you need to improve.

This guide is challenging - it asks you to do more, not less.

It starts with the fundamental skills, but does not end there; it also includes the *most advanced content* that many other prep books ignore. As the average GMAT score required to gain admission to top business schools continues to rise, this guide, together with the 6 computer adaptive online practice exams and bonus question bank included with your purchase, provides test-takers with the depth and volume of advanced material essential for achieving the highest scores, given the GMAT's computer adaptive format.

This guide is ambitious - developing mastery is its goal.

Developed by Manhattan GMAT's staff of REAL teachers (all of whom have 99th percentile official GMAT scores), our ambitious curriculum seeks to provide test-takers of all levels with an in-depth and carefully tailored approach that enables our students to achieve mastery. If you are looking to learn more than just the "process of elimination" and if you want to develop skills, strategies, and a confident approach to any problem that you may see on the GMAT, then our sophisticated preparation guides are the tools to get you there.

HOW TO ACCESS YOUR ONLINE RESOURCES

Please read this entire page of information, all the way down to the bottom of the page! This page describes WHAT online resources are included with the purchase of this book and HOW to access these resources.

[**If you are a registered Manhattan GMAT student** and have received this book as part of your course materials, you have AUTOMATIC access to ALL of our online resources. This includes all simulated practice exams, question banks, and online updates to this book. To access these resources, follow the instructions in the Welcome Guide provided to you at the start of your program. Do NOT follow the instructions below.]

If you have purchased this book, your purchase includes 1 YEAR OF ONLINE ACCESS to the following:

6 Computer Adaptive Online Practice Exams

Bonus Online Question Bank for EQUATIONS, INEQUALITIES, & VIC'S

Online Updates to the Content in this Book

The 6 full-length computer adaptive practice exams included with the purchase of this book are delivered online using Manhattan GMAT's proprietary computer adaptive online test engine. The exams adapt to your ability level by drawing from a bank of more than 1200 unique questions of varying difficulty levels written by Manhattan GMAT's expert instructors, all of whom have scored in the 99th percentile on the Official GMAT. At the end of each exam you will receive a score, an analysis of your results, and the opportunity to review detailed explanations for each question. You may choose to take the exams timed or untimed.

The Bonus Online Question Bank for Equations, Inequalities, & VIC's consists of 25 extra practice questions (with detailed explanations) that test the variety of Equation, Inequality, & VIC concepts and skills covered in this book. These questions provide you with extra practice *beyond* the problem sets contained in this book. You may use our online timer to practice your pacing by setting time limits for each question in the bank.

The content presented in this book is updated periodically to ensure that it reflects the GMAT's most current trends. You may view all updates, including any known errors or changes, upon registering for online access.

Important Note: The 6 computer adaptive online exams included with the purchase of this book are the SAME exams that you receive upon purchasing ANY book in Manhattan GMAT's 8 Book Preparation Series. On the other hand, the Bonus Online Question Bank for EQUATIONS, INEQUALITIES, & VIC'S is a unique resource that you receive ONLY with the purchase of this specific title.

To access the online resources listed above, you will need this book in front of you and you will need to register your information online. This book includes access to the above resources for ONE PERSON ONLY.

To register and start using your online resources, please go online to the following URL:

http://www.manhattangmat.com/access.cfm (Double check that you have typed this in accurately!)

Your one-year of online access begins on the day that you register at the above URL. You only need to register your product ONCE at the above URL. To use your online resources any time AFTER you have completed the registration process, please login to the following URL:

http://www.manhattangmat.com/practicecenter.cfm

TABLE OF CONTENTS

g

Chapter 1

of

EQUATIONS, INEQUALITIES, & VIC's

BASIC EQUATIONS

In This Chapter . . .

- Solving 1-Variable Equations
- Simultaneous Equations: Solving by Substitution
- Simultaneous Equations: Solving by Combination
- Simultaneous Equations: 3 Equations
- Mismatch Problems
- Combo Problems: Manipulations
- Absolute Value Equations

BASIC EQUATIONS

Algebra is one of the major math topics tested on the GMAT. Your ability to solve equations is, therefore, an essential component to success.

Basic GMAT equations are those that DO NOT involve exponents. There are several different types of BASIC equations that the GMAT expects you to solve:

1) An equation with 1 variable
2) Simultaneous equations with 2 or 3 variables
3) Mismatch Equations
4) Combos
5) Equations with absolute value

To solve basic equations, remember that whatever you do to one side, you must also do to the other side.

Several of the preceding basic equation types probably look familiar to you. Others—particularly Mismatch Equations and Combos—are unique GMAT favorites that run counter to some of the rules you may have learned in high-school algebra. Being attuned to the particular subtleties of GMAT equations can be the difference between an average score and an excellent one.

Solving 1-Variable Equations

Equations with one variable should be familiar to you from previous encounters with algebra. In order to solve 1-variable equations, simply isolate the variable on one side of the equation. In doing so, make sure you perform identical operations to both sides of the equation. Here are three examples:

$3x + 5 = 26$	Subtract 5 from both sides.
$3x = 21$	Divide both sides by 3.
$x = 7$	

$w = 17w - 1$	Subtract w from both sides.
$0 = 16w - 1$	Add 1 to both sides.
$1 = 16w$	Divide both sides by 16.
$\frac{1}{16} = w$	

$\frac{p}{9} + 3 = 5$	Subtract 3 from both sides.
$\frac{p}{9} = 2$	Multiply both sides by 9.
$p = 18$	

Simultaneous Equations: Solving by Substitution

Sometimes the GMAT asks you to solve a system of equations with more than one variable. You might be given two equations with two variables, or three equations with three variables. In either case, there are two primary ways of solving simultaneous equations: by substitution, or by combination.

Solve the following for x and y.

$$x + y = 9$$
$$2x = 5y + 4$$

Use substitution whenever one variable can be easily expressed in terms of the other.

1. Solve the first equation for x.

$$x + y = 9$$
$$x = 9 - y$$

2. Substitute this solution into the second equation wherever x appears.

$$2x = 5y + 4$$
$$2(9 - y) = 5y + 4$$

3. Solve the second equation for y.

$$2(9 - y) = 5y + 4$$
$$18 - 2y = 5y + 4$$
$$14 = 7y$$
$$2 = y$$

4. Substitute your solution for y into the first equation in order to solve for x.

$$x + y = 9$$
$$x + 2 = 9$$
$$x = 7$$

Simultaneous Equations: Solving by Combination

Alternatively, you can solve simultaneous equations by combination. In this method, add or subtract the two equations to eliminate one of the variables.

Use combination whenever it is easy to manipulate the equations so that the coefficients for one variable are the SAME or OPPOSITE.

Solve the following for x and y.

$$x + y = 9$$
$$2x = 5y + 4$$

1. Line up the terms of the equations.

$$x + y = 9$$
$$2x - 5y = 4$$

2. If you plan to add the equations, multiply one or both of the equations so that the coefficient of a variable in one equation is the OPPOSITE of that variable's coefficient in the other equation. If you plan to subtract them, multiply one or both of the equations so that the coefficient of a variable in one equation is the SAME as that variable's coefficient in the other equation.

$$-2(x + y = 9) \rightarrow -2x - 2y = -18$$
$$2x - 5y = 4 \rightarrow 2x - 5y = 4$$

Note that the x coefficients are now opposites.

3. Add the equations to eliminate one of the variables.

$$\begin{array}{r} -2x - 2y = -18 \\ +\ \ 2x - 5y = \ \ \ 4 \\ \hline -7y = -14 \end{array}$$

4. Solve the resulting equation for the unknown variable.

$$-7y = -14$$
$$y = 2$$

5. Substitute into one of the original equations to solve for the second variable.

$$x + y = 9$$
$$x + 2 = 9$$
$$x = 7$$

Simultaneous Equations: 3 Equations

The procedure for solving a system of three equations with 3 variables is exactly the same. You can use substitution or combination. This example uses both:

Solve the following for w, x, and y.

$$x + w = y$$
$$2y + w = 3x - 2$$
$$13 - 2w = x + y$$

Solve three simultaneous equations step-by-step. Keep careful track of your work to avoid careless errors.

1. The first equation is already solved for y.

$$y = x + w$$

2. Substitute for y in the second and third equations.

Substitute for y in the second equation:

$$2(x + w) + w = 3x - 2$$
$$2x + 2w + w = 3x - 2$$
$$-x + 3w = -2$$

Substitute for y in the third equation:

$$13 - 2w = x + (x + w)$$
$$13 - 2w = 2x + w$$
$$3w + 2x = 13$$

3. Multiply the first of the resulting two-variable equations by (-1) and combine them with addition.

$$\begin{array}{r} x - 3w = 2 \\ +\quad 2x + 3w = 13 \\ \hline 3x = 15 \end{array}$$

Therefore, $x = 5$

4. Use your solution for x to determine solutions for the other two variables.

$$3w + 2x = 13$$
$$3w + 10 = 13$$
$$3w = 3$$
$$w = 1$$

$$y = x + w$$
$$y = 5 + 1$$
$$y = 6$$

Mismatch Problems

Consider the following rule, which you might have learned in a basic algebra course: If you are trying to solve for 2 different variables, you need 2 equations. If you are trying to solve for 3 different variables, you need 3 equations, etc. The GMAT loves to trick you by taking advantage of your faith in this easily misapplied rule.

MISMATCH problems, which are particularly common on the Data Sufficiency portion of the test, are those in which the number of unknown variables does NOT correspond to the number of given equations. Do not try to apply that old rule you learned in high-school algebra, because it is not applicable for many GMAT equation problems. All MISMATCH problems must be solved on a case-by-case basis.

Don't assume that the number of equations must be equal to the number of variables.

Solve for x given the following two equations:

$$\textbf{(1)}\ \frac{3x}{2y+2z}=3 \qquad \textbf{(2)}\ y+z=5$$

It is tempting to say that these two equations are not sufficient to solve for x, since there are 3 variables and only 2 equations. However, note that the question does NOT ask you to solve for all three variables. It only asks you to solve for x, which IS possible:

First, solve the first equation for x, in terms of y and z:

$$\frac{3x}{2y+2z}=3$$
$$3x=3(2y+2z)$$
$$x=2y+2z$$
$$x=2(y+z)$$

Then, notice that the second equation gives us a value for $y + z$, which we can substitute into the first equation in order to solve for x.

$$x=2(y+z)$$
$$x=2(5)$$
$$x=10$$

Now consider an example in which 2 equations with 2 unknowns are actually NOT sufficient to solve a problem:

Solve for x given the following two equations:

$$\textbf{(1)}\ x+12y=15 \qquad \textbf{(2)}\ \frac{x}{3}+4y=5$$

It is tempting to say that these 2 equations are surely sufficient to solve for x, since there are 2 equations and only 2 variables. However, notice that both equations actually express the same relationship between x and y. That is, they are the same equation!

If you divide the first equation by 3, you will obtain the second equation. Alternatively, if you multiply the second equation by 3, you will obtain the first equation.

Thus, we have only one equation with 2 variables. We do NOT have sufficient information to solve for x.

Combo Problems: Manipulations

The GMAT often asks you to solve for a combination of variables, called COMBO problems. For example, a question might ask, what is the value of $x + y$?

In these cases, since you are not asked to solve for one specific variable, you should generally NOT try to solve for the individual variables right away. Instead, you should try to manipulate the given equation(s) so that the COMBO is isolated on one side of the equation. Only try to solve for the individual variables after you've exhausted all other avenues.

To solve for a variable combo, isolate the combo on one side of the equation.

There are 3 easy manipulations that are the key to solving most COMBO problems. You can use the acronym **MUD** to remember them.

M: Multiply or divide the whole equation by a certain number.
U: Unsquare or square both sides.
D: Distribute and factor.

Here are two examples, each of which uses one or more of the manipulations above:

If $x = \frac{7 - y}{2}$, what is $2x + y$?

$$x = \frac{7 - y}{2}$$

$$2x = 7 - y$$
$$2x + y = 7$$

Here, getting rid of the denominator by multiplying both sides of the equation by 2 is the key to isolating the combo on one side of the equation.

If $\sqrt{2t + r} = 5$, what is $3r + 6t$?

$$\left(\sqrt{2t + r}\right)^2 = 5^2$$
$$2t + r = 25$$
$$6t + 3r = 75$$

Here, getting rid of the square root by squaring both sides of the equation is the first step. Then, multiplying the whole equation by 3 forms the combo in question.

Absolute Value Equations

Equations that involve absolute value generally have TWO SOLUTIONS. It is important to consider this rule when thinking about GMAT questions that involve absolute value. The following three-step method should be used for solving equations that include a variable inside an absolute value sign.

Solve for y, given that $|y| + 7 = 15$.

1. Isolate the expression within the absolute value brackets.

$$|y| + 7 = 15$$
$$|y| = 8$$

2. Remove the absolute value brackets and solve the equation for 2 cases. For the first case, assume the variable inside the absolute value brackets is positive. For the second case, assume the variable inside the absolute value brackets is negative.

CASE 1: $y > 0$
$y = 8$

CASE 2: $y < 0$
$y = -8$

3. Check to see if each solution is valid.

In case 1, we set the following condition: $y > 0$. The solution, $y = 8$ meets this condition (since 8 is greater than 0).

In case 2, we set the following condition: $y < 0$. The solution, $y = -8$ meets this condition (since -8 is less than 0).

Thus, both solutions are valid. A solution would not be valid if it violated the pre-set condition. For example, if we set a condition that $y > 0$ and came to a solution that $y = -2$, then we could discard this solution (since, in effect, we would be saying that if y is positive, it must be -2, a statement that does not make logical sense).

Equations that involve a variable inside the absolute value brackets generally have two solutions: one for when the expression inside the brackets is positive, and one for when the expression inside the brackets is negative. The exception to this general rule is when the solution is zero.

Absolute value equations must be solved for both the positive and negative cases.

Absolute Value Equations with Variable Expressions

Apply the same 3-step method when there is a variable *expression* within the absolute value brackets.

Solve for w, given that $12 + |w - 4| = 30$.

Again, isolate the expression within the absolute value brackets and consider both cases.

1. $|w - 4| = 18$

2. CASE 1: $w - 4 > 0$
   $w - 4 = 18$
   $w = 22$

   CASE 2: $w - 4 < 0$
   $w - 4 = -18$
   $w = -14$

3. The first solution, $w = 22$, is valid because it meets the preset condition that $w - 4 > 0$ (since $22 - 4 = 18$, which is greater than 0.)

   The second solution, $w = -14$, is valid because it meets the preset condition that $w - 4 < 0$ (since $-14 - 4 = -18$, which is less than 0.)

Don't forget to determine the validity of each of your solutions to absolute value equations by checking each solution against any relevant preset conditions.

Problem Set

For problems #1-6, solve for all unknowns.

1. $\frac{3x-6}{5}=x-6$

2. $\frac{x+2}{4+x}=\frac{5}{9}$

3. $22-|y+14|=20$

4. $|6+x|=2x+1$

5. $y=2x+9$ and $7x+3y=-51$

6. $3x+y+7z=15$ and $x+2y+(-4z)=-1$ and $5x+2y+8z=19$

For problems #7-10, determine whether it is *possible* to solve for x using the given equations. (Do not solve.)

7. $\frac{\sqrt{x}}{6a}=T$ and $\frac{Ta}{4}=14$

8. $3x+2a=8$ and $9x+6a=24$

9. $3a+2b+x=8$ and $12a+8b+2x=4$

10. $4a+7b+9x=17$ and $3a+3b+3x=3$ and $9a+\frac{b}{4}+\frac{x}{9}=9$

For problems #11-15, solve for the specified expression.

11. Given that $\frac{x+y}{3}=17$, what is $x+y$?

12. Given that $\frac{a+b}{b}=21$, what is $\frac{a}{b}$?

13. Given that $10x+10y=x+y+81$, what is $x+y$?

14. Given that $5x+9y+4=2x+3y+31$, what is $x+2y$?

15. Given that $\frac{zx+zy}{9}=4$ and $x+y=3$, what is z?

1. $\mathbf{x = 12}$:

$$\frac{3x-6}{5} = x - 6$$

$$3x - 6 = 5(x - 6)$$
$$3x - 6 = 5x - 30$$
$$24 = 2x$$
$$12 = x$$

Solve by multiplying both sides by 5 to eliminate the denominator. Then, distribute and isolate the variable on the left side.

2. $\mathbf{x = \frac{1}{2}}$:

$$\frac{x+2}{4+x} = \frac{5}{9}$$

$$9(x + 2) = 5(4 + x)$$
$$9x + 18 = 20 + 5x$$
$$4x = 2$$
$$x = \frac{1}{2}$$

Cross-multiply to eliminate the denominators. Then, distribute and solve.

3. $\mathbf{y = \{-16, -12\}}$:

$$22 - |y + 14| = 20$$
$$|y + 14| = 2$$

First, isolate the expression within the absolute value brackets. Then, solve for two cases, one in which the expression is positive and one in which it is negative.

Case 1: $y + 14 > 0$
$y + 14 = 2$
$y = -12$

Case 2: $y + 14 < 0$
$y + 14 = -2$
$y = -16$

4. $\mathbf{x = 5}$: Solve this equation for two cases, one in which the expression is positive and one in which it is negative. Notice that the solution arrived at in Case 2 is not valid, since it violates the condition that x be less than -6.

Case 1: When $6 + x$ is positive (meaning $x > -6$)

$$|6 + x| = 2x + 1$$
$$6 + x = 2x + 1$$
$$5 = x$$

Case 2: When $6 + x$ is negative (meaning $x < -6$)

$$6 + x = -(2x + 1)$$
$$6 + x = -2x - 1$$
$$3x = -7$$
$$x = -\frac{7}{3} \quad \textit{not valid!}$$

5. $\mathbf{x = -6; y = -3}$:

$$y = 2x + 9 \qquad 7x + 3y = -51$$
$$7x + 3(2x + 9) = -51$$
$$7x + 6x + 27 = -51$$
$$13x + 27 = -51$$
$$13x = -78$$
$$x = -6$$
$$y = 2x + 9 = 2(-6) + 9 = -3$$

Solve this system by substitution. Substitute the value given for y in the first equation into the second equation. Then, distribute, combine like terms, and solve. Once you get a value for x, substitute it back into the first equation to obtain the value of y.

6. $\mathbf{x = -1; y = 4; z = 2}$:

$3x + y + 7z = 15$ and $x + 2y + (-4z) = -1$ and $5x + 2y + 8z = 19$

$$\begin{array}{lcr} (2)[3x + y + 7z = 15] & \rightarrow & 6x + 2y + 14z = 30 \\ (-1)[(x + 2y - 4z) = -1] & \rightarrow + & \underline{-x - 2y + 4z = 1} \\ & & 5x \quad + 18z = 31 \end{array}$$

First, combine equations (1) and (2) by addition. Multiply them each by a single number as shown to eliminate one variable.

$$\begin{array}{lcr} (-1)[x + 2y + (-4z)] = -1 & \rightarrow & -x - 2y + 4z = 1 \\ & + & \underline{5x + 2y + 8z = 19} \\ & & 4x \quad + 12z = 20 \end{array}$$

Then, combine equations (2) and (3). In this case, you can eliminate a variable by altering only one of the equations.

$$\begin{array}{lcr} (4)[5x + 18z = 31] & \rightarrow & 20x + 72z = 124 \\ (-5)[4x + 12z = 20] & \rightarrow + & \underline{-20x - 60z = -100} \\ & & 12z = 24 \\ & & z = 2 \end{array}$$

Then, combine the resulting 2-variable equations to isolate the variable z.

$$4x + 12z = 20 \qquad 3x + y + 7z = 15$$
$$4x + 12(2) = 20 \qquad 3(-1) + y + 7(2) = 15$$
$$4x + 24 = 20 \qquad -3 + y + 14 = 15$$
$$4x = -4 \qquad 11 + y = 15$$
$$x = -1 \qquad y = 4$$

Finally, plug the known value for z into a 2-variable equation to get the value of x. Then, plug the known values of z and x into any of the original equations to get the value of y.

7. **YES:** This problem contains 3 variables and 2 equations. However, this is not enough to conclude that you cannot solve for x. You must check to see if you can solve by isolating a combination of variables, as shown below:

$$\frac{\sqrt{x}}{6a} = T \text{ and } \frac{Ta}{4} = 14$$

$$x = (6Ta)^2 \text{ and } Ta = 56$$
$$x = (6 \times 56)^2$$

We can find a value for x.

8. **NO:** This problem contains 2 variables and 2 equations. However, this is not enough to conclude that you can solve for x. If one equation is merely a multiple of the other one, then you will not have a unique solution for x. In this case, the second equation is merely 3 times the first. Therefore, they cannot be combined to find the value of x.

9. **YES:** This problem contains 3 variables and 2 equations. However, this is not enough to conclude that you cannot solve for x. You must check to see if you can solve by eliminating all the variables but x, as shown below:

$$(4)[3a + 2b + x = 8] \rightarrow \quad 12a + 8b + 4x = 32$$
$$- \quad 12a + 8b + 2x = 4$$
$$2x = 28$$
$$x = 14$$

10. **YES:** This problem contains 3 variables and 3 equations. However, this is not enough to conclude that you can solve for x. You must check to make sure that they are 3 unique equations. In fact, they are, as you can see by multiplying each equation to establish a common coefficient in the x term:

$$(3)[3a + 3b + 3x = 3] \rightarrow \quad 9a + 9b + 9x = 9$$
$$(81)[9a + \frac{b}{4} + \frac{x}{9} = 9] \rightarrow \quad 729a + 20.25b + 9x = 9$$
$$4a + 7b + 9x = 17$$

If required to solve for x, you could do so by systematically combining these equations to isolate individual variables. However, since you are only asked to decide whether or not you could solve for x, it is sufficient to establish that these are 3 distinct equations.

11. $\boldsymbol{x + y = 51}$**:**

$$\frac{x+y}{3} = 17$$

$$x + y = 51$$

12. $\boldsymbol{\frac{a}{b} = 20}$**:**

$$\frac{a+b}{b} = 21$$

$$\frac{a}{b} + 1 = 21$$

$$\frac{a}{b} = 20$$

13. $\boldsymbol{x + y = 9}$**:**

$$10x + 10y = x + y + 81$$
$$9x + 9y = 81$$
$$x + y = 9$$

14. $\boldsymbol{x + 2y = 9}$**:**

$$5x + 9y + 4 = 2x + 3y + 31$$
$$3x + 6y = 27$$
$$3(x + 2y) = 27$$
$$x + 2y = 9$$

15. $\boldsymbol{z = 12}$**:**

$$\frac{zx + zy}{9} = 4 \qquad x + y = 3$$

$$z(x + y) = 36$$
$$3z = 36$$
$$z = 12$$

Chapter 2

of

EQUATIONS, INEQUALITIES, & VIC's

EXPONENTIAL EQUATIONS

In This Chapter . . .

- Even Exponent Equations: 2 Solutions
- Odd Exponents: 1 Solution
- Same Base or Same Exponent
- Eliminating Roots: Square Both Sides

EXPONENTIAL EQUATIONS

The GMAT tests more than your knowledge of basic equations. In fact, the GMAT complicates equations by including exponents or roots with the unknown variables.

Exponential equations take various forms. Here are some examples:

$x^3 = -125$ $\quad$ $y^2 + 3 = x$ $\quad$ $\sqrt{x} + 15 = 21$

There are two keys to achieving success with exponential equations:

1) Know the RULES for exponents and roots. You will recall that these rules were covered in the Number Properties Strategy Pack. It is essential to know these rules by heart. In particular, you should review the rules for combining exponential expressions.

Even exponents are dangerous! They hide the sign of the base.

2) Remember that EVEN EXPONENTS are DANGEROUS because they hide the sign of the base. In general, equations with even exponents have 2 solutions.

Even Exponent Equations: 2 Solutions

Recall from the rules of exponents that **even exponents are dangerous** in the hands of the GMAT test writers.

Why are they dangerous? Even exponents hide the sign of the base. As a result, equations that involve variables with even exponents can have both a positive and a negative solution.

Exponential equations with even exponents must be carefully analyzed, because they often have 2 solutions. Here are some examples:

$x^2 = 9$ $\quad$ Here, x has two solutions: 3 and -3.

$a^2 - 5 = 12$ $\quad$ By adding 5 to both sides, we can rewrite this equation as $a^2 = 17$. Thus, a has two solutions: $\sqrt{17}$ and $-\sqrt{17}$.

Note that not all equations with even exponents have 2 solutions. For example:

$x^2 + 3 = 3$ $\quad$ By subtracting 3 from both sides, we can rewrite this equation as $x^2 = 0$, so x has only one solution: 0.

Also, BE ON THE LOOKOUT for even exponential equations in the form $x^2 + k = 0$, where k is any positive integer. These equations have NO REAL solutions:

$x^2 + 9 = 0$ $\quad$ Squaring can never produce a negative number!
$x^2 = -9$

Odd Exponents: 1 Solution

Equations that involve only odd exponents or cube roots have only 1 solution:

$x^3 = -125$ Here, x has only 1 solution: -5. You can see that $(-5)(-5)(-5) = -125$. This will not work with positive 5.

$243 = y^5$ Here, y has only 1 solution: 3. You can see that $(3)(3)(3)(3)(3) = 243$. This will not work with negative 3.

If an equation includes some variables with odd exponents and some variables with even exponents, treat it as dangerous, as it is likely to have 2 solutions. Any even exponents in an equation make it dangerous.

Rewrite exponential equations so they have either the same base or the same exponent.

Same Base or Same Exponent

In problems that involve exponential expressions on BOTH sides of the equation, it is imperative to REWRITE the bases so that either the same base or the same exponent appears on both sides of the exponential equation. Once you do this, you can eliminate the bases or the exponents and rewrite the remainder as an equation.

Solve the following equation for w: $(4^w)^3 = 32^{w-1}$

1. Rewrite the bases so that the same base appears on both sides of the equation. Right now, the left side has a base of 4 and the right side has a base of 32. Notice that both 4 and 32 can be expressed as powers of 2.

   We can rewrite 4 as 2^2, and we can rewrite 32 as 2^5.

2. Plug the rewritten bases into the original equation.

   $(4^w)^3 = 32^{w-1}$
   $((2^2)^w)^3 = (2^5)^{w-1}$

3. Simplify the equation using the rules of exponents.

   $((2^2)^w)^3 = (2^5)^{w-1}$
   $2^{6w} = 2^{5w-5}$

4. Eliminate the identical bases, rewrite the exponents as an equation, and solve.

   $6w = 5w - 5$
   $w = -5$

Eliminating Roots: Square Both Sides

The most effective way to solve problems that involve variables underneath radical symbols (variable square roots) is to square both sides of the equation.

Solve the following equation for s: $\sqrt{s - 12} = 7$

$\sqrt{s-12} = 7$
$s - 12 = 49$
$s = 61$

Squaring both sides of the equation eliminates the radical symbol and allows us to solve for s more easily.

When eliminating radicals, you may sometimes need to square both sides of the equation more than once.

Given that $\sqrt{3b - 8} = \sqrt{12 - b}$, what is b?

$\sqrt{3b-8} = \sqrt{12-b}$
$3b - 8 = 12 - b$
$4b - 8 = 12$
$4b = 20$
$b = 5$

Squaring both sides of the equation eliminates both radical symbols and allows us to solve for b more easily.

After you've solved for the variable, check that the solution works in the original equation. Squaring both sides can actually introduce an extraneous solution.

Remember also that the written square root symbol only works over positive numbers (or zero) and only yields positive numbers (or zero). The square root of a negative number is not defined in the real number system.

$\sqrt{x} = 5$ means $x = 25$

At the same time, there are two numbers whose square is 25: 5 and -5.

For equations that involve cube roots, solve by cubing both sides of the equation:

Solve the following equation for y: $-3 = \sqrt[3]{y-8}$

$-3 = \sqrt[3]{y-8}$
$-27 = y - 8$
$-19 = y$

Cubing both sides of the equation eliminates the radical.

Problem Set

1. Given that $\sqrt{t+8} = 6$, what is t?

2. Given that $\sqrt{m+4} = \sqrt{2m-11}$, what is m?

3. Given that $\sqrt{\sqrt{2x}+23} = 5$, what is x?

4. If u is a positive integer, which of the following could be a negative number?
(A) $u^7 - u^6$ (B) $u^3 + u^4 + u^5$ (C) u^{-9} (D) $u^{-13} + u^{13}$ (E) $u^3 - u^8$

5. Simplify: $\left(\frac{x^{21}}{x^3}\right)^7$

6. Given that $2x + y = -3$ and $x^2 = y + 3$, solve for x and y.

7. If $x^4 = 81$, $z^3 = -125$, and $d^2 = 4$, what is the smallest possible value of $x + z + d$?

8. Given that k and m are positive integers and that $(x^4)(x^8) = (x^k)^m$ and $m = k + 1$, solve for k and m.

9. Given that $(3^k)^4 = 27$, what is k?

10. Given that $x + y = 13$ and $\sqrt{y} = x - 1$, solve for the *positive* values of x and y.

For problems #11-13, given that x is an integer greater than 1, determine whether each of the following expressions can be an integer.

11. $x^7 + x^{-7}$

12. $x^{\frac{1}{4}} + x^{\frac{1}{2}}$

13. $x^{\frac{1}{3}} + x^0 + x^5$

For problems #14 and #15, use the following information: At a shoe store's fairly drastic sale, prices are defined as the square root of ten dollars more than the original dollar price.

14. If a customer buys a pair of discounted shoes and gets change for a twenty-dollar bill, what is the highest possible original cost of the shoes (in whole dollars)?

15. A salesman sells a shoehorn for the sale price of 80 dollars, even though this item was not supposed to have been part of the storewide sale. When the shoe store owner tries to buy it back, the crafty purchaser holds out for twice the original price. How much of a profit does the crafty buyer make?

1. **28:**

$$\sqrt{t+8} = 6$$
$$t + 8 = 36$$
$$t = 28$$

Square both sides to eliminate the radical. Then, solve for t.

2. **15:**

$$\sqrt{m+4} = \sqrt{2m - 11}$$
$$m + 4 = 2m - 11$$
$$15 = m$$

Square both sides to eliminate the radicals. Then, solve for m.

3. **x = 2:**

$$\sqrt{\sqrt{2x} + 23} = 5$$
$$\sqrt{2x} + 23 = 25$$
$$\sqrt{2x} = 2$$
$$2x = 4$$
$$x = 2$$

Square both sides; this eliminates the larger radical sign on the left side of the equation. Then, subtract 23 from both sides to isolate the variable. Square both sides again to eliminate the radical. Finally, divide both sides by 2 to find the value of x.

4. **(E):** If u is a positive integer, $u^8 > u^3$. Therefore, $u^3 - u^8 < 0$.

5. **x^{126}:**

$$\left(\frac{x^{21}}{x^3}\right)^7 = (x^{21-3})^7 = (x^{18})^7 = x^{18\cdot 7} = x^{126}$$

6. **$x = 0, y = -3$ OR $x = -2, y = 1$ (both solutions are possible)**

$$2x + y = -3 \qquad\qquad x^2 = y + 3$$
$$y = -2x - 3 \qquad\qquad y = x^2 - 3$$

$$-2x - 3 = x^2 - 3$$
$$x^2 + 2x = 0$$
$$x(x+2) = 0$$
$$x = 0 \quad \text{OR} \quad x + 2 = 0$$
$$x = -2$$

Rearrange each equation so that it expresses y in terms of x. Then, set the right sides of both equations equal to each other and solve for x. Substitute each value of x into either equation to find the corresponding value for y.

If $x = 0$, then $y = x^2 - 3 = (0)^2 - 3 = -3$

If $x = -2$, then $y = x^2 - 3 = (-2)^2 - 3 = 1$

7. **−10:**

If $x^4 = 81$, $x = \{-3, 3\}$.
If $z^3 = -125$, $z = -5$.
If $d^2 = 4$, $d = \{-2, 2\}$.

To find the smallest possible value of $x + z + d$, select the smallest value for each variable:

$x = -3 \qquad y = -5 \qquad z = -2$

$x + z + d = -10$

8. **$k = 3$, $m = 4$:**

$(x^4)(x^8) = (x^k)^m$ and $m = k + 1$

$$x^{12} = x^{km}$$
$$km = 12$$
$$k(k + 1) = 12$$
$$k^2 + k = 12$$
$$k^2 + k - 12 = 0$$
$$(k - 3)(k + 4) = 0$$
$$k - 3 = 0 \quad \text{OR} \quad k + 4 = 0$$
$$k = 3 \qquad\qquad k = -4$$

Since we know that k and m are both positive integers, k must be equal to 3. Therefore, $m = k + 1 = 4$.

9. **$k = \frac{3}{4}$:**

$$(3^k)^4 = 27$$
$$3^{4k} = 3^3$$
$$4k = 3$$
$$k = \frac{3}{4}$$

10. **$x = 4$, $y = 9$:**

$x + y = 13 \qquad \sqrt{y} = x - 1$

$y = 13 - x \qquad y = (x - 1)(x - 1)$

$\qquad\qquad y = x^2 - 2x + 1$

Solve both equations for y in terms of x. Then, set the right side of each equation equal to each other. Solve the quadratic equation by factoring. Since we are asked only for the positive values, discard −3. Then, substitute the remaining value for x, 4, into either equation to find the corresponding value for y.

$$13 - x = x^2 - 2x + 1$$
$$x^2 - x - 12 = 0$$
$$(x - 4)(x + 3) = 0$$
$$x - 4 = 0 \quad \text{OR} \quad x + 3 = 0$$
$$x = 4 \qquad\qquad x = -3$$

$$y = 13 - x = 13 - 4 = 9$$

11. **NO:** The only value of x for which $x^7 + x^{-7}$ can be an integer is 1.
$(x^7 + x^{-7} = 1^7 + 1^{-7} = 1 + 1 = 2)$
However, we know that x must be an integer greater than 1.

12. **YES:** If $x = 16$, $x^{\frac{1}{4}} + x^{\frac{1}{2}} = \sqrt[4]{16} + \sqrt{16} = 2 + 4 = 6$. If x is any number with an integer fourth root, this sum will be an integer.

13. **YES:** If $x = 8$, $x^{\frac{1}{3}} + x^0 + x^5 = \sqrt[3]{8} + 8^0 + 8^5 = 2 + 1 + 32{,}768 = 32{,}771$. If x is any perfect cube, this sum will be an integer.

14. **$389:**

$$\begin{aligned} \sqrt{x + 10} &< 20 \\ x + 10 &< 400 \\ x &< 390 \end{aligned}$$

The highest possible price less than $390 is $389.

15. **$12,700:**

Sale price = $80

$$\begin{aligned} \sqrt{x + 10} &= 80 \\ x + 10 &= 6{,}400 \\ x &= 6{,}390 \end{aligned}$$

Original price = $6,390
Buyback price = 2($6,390) = $12,780
Profit: $12,780 − $80 = $12,700

Chapter 3

of

EQUATIONS, INEQUALITIES, & VIC's

QUADRATIC EQUATIONS

In This Chapter . . .

- Factoring Quadratic Equations
- Disguised Quadratics
- Going in Reverse: Use FOIL
- Using FOIL with Square Roots
- 1-Solution Quadratics
- Zero in the Denominator: Undefined
- The 3 Favorite Forms

QUADRATIC EQUATIONS

One special type of even exponent equation is called the quadratic equation. Here are some examples of quadratic equations:

$x^2 + 3x + 8 = 12$ $\quad$ $w^2 - 16w + 1 = 0$ $\quad$ $2y^2 - y + 5 = 8$

Quadratic equations are equations with one unknown and two defining components:
(1) a variable term raised to the second power
(2) a variable term raised to the first power

Although the standard form of the quadratic equation is like the examples above, here are other ways of writing quadratics:

$x^2 = 3x + 4$ $\quad$ $a = 5a^2$ $\quad$ $6 - b = 7b^2$

All quadratic equations on the GMAT can be factored. You will never need to use the quadratic formula.

Notice that the preceding examples each contain a variable raised to the first power and a variable raised to the second power.

Like other even exponent equations, quadratic equations generally have 2 solutions.

Factoring Quadratic Equations

The following example illustrates the process for solving quadratic equations:

What is x, given that $x^2 + 3x + 8 = 12$?

1. Move all the terms to the left side of the equation, combine them, and put them in the form $ax^2 + bx + c$ (where a, b, and c are integers). The right side of the equation should be set to 0. (Usually, this process makes the x^2 term positive. If not, move all the terms to the right side of the equation instead.)

$x^2 + 3x + 8 = 12$ Subtracting 12 from both sides of the equation puts all the
$x^2 + 3x - 4 = 0$ terms on the left side, and sets the right side to 0.

2. Factor the equation. In order to factor, you generally need to think about two terms in the equation. Assuming that $a = 1$ (which is almost always the case on the GMAT), the two terms you should focus on are b and c. The trick to factoring is to find two integers whose product equals c and whose sum equals b.

In the equation $x^2 + 3x - 4 = 0$, we can see that $b = 3$ and $c = -4$. In order to factor this equation, we need to find two integers whose product is -4 and whose sum is 3. The only two integers that work are 4 and -1, since $4(-1) = -4$ and $4 + (-1) = 3$.

3. Rewrite the equation in the form $(x + ?)(x + ?)$, where the question marks represent the two integers you solved for in the previous step.

$x^2 + 3x - 4 = 0$
$(x + 4)(x - 1) = 0$

4. The equation is now expressed as a product of two factors. **Since this product equals 0, one or both of the factors must be 0.** Set each factor independently to 0 and solve for x.

$x + 4 = 0$	OR	$x - 1 = 0$	The two solutions for x have the opposite signs of the integers we found in step three.
$x = -4$		$x = 1$	

Beware of disguised quadratics!

Disguised Quadratics

The GMAT will often attempt to disguise quadratic equations by putting them in forms that don't quite look like the traditional form of $ax^2 + bx + c = 0$.

Here is the most common "disguised" form for a quadratic:

$3w^2 = 6w$

This is certainly a quadratic equation. However, it is very tempting to try to solve this equation without thinking of it as a quadratic. This classic mistake looks like this:

$3w^2 = 6w$
$3w = 6$
$w = 2$

Dividing both sides by w and then dividing both sides by 3 yields the solution $w = 2$

In solving this equation without factoring it like a quadratic, we have missed one of the solutions! Let's now solve it by factoring it as a quadratic equation:

$3w^2 = 6w$
$3w^2 - 6w = 0$
$w(3w - 6) = 0$

Setting both factors equal to 0 yields the following solutions:

$w = 0$ OR

$3w - 6 = 0$
$3w = 6$
$w = 2$

In recognizing that $3w^2 = 6w$ is a disguised quadratic, we have found both solutions instead of accidentally missing one!

Here is another example of a disguised quadratic:

Solve for b, given that $\frac{36}{b} = b - 5$.

Although at first glance, this does not look like a quadratic equation, once we begin solving it we must recognize that it is a quadratic!

$\frac{36}{b} = b - 5$

$36 = b^2 - 5b$

We start by multiplying both sides of the equation by b. After we do this, we should recognize the components of a quadratic equation.

Manipulations can help you uncover disguised quadratics.

Now we should treat this as a quadratic equation and solve it by factoring:

$36 = b^2 - 5b$

$b^2 - 5b - 36 = 0$

$(b - 9)(b + 4) = 0$ Thus, $b = 9$ or $b = -4$.

Some quadratics are hidden within more complex equations. For example:

The equation $x^3 + x^2 - 2x = 0$ can be rewritten as follows:

$x^3 + x^2 - 2x = 0$

$x(x^2 + x - 2) = 0$

In factoring out an x from each term, we are left with the product of x and the quadratic $x^2 + x - 2$.

Now we can factor the hidden quadratic:

$x(x^2 + x - 2) = 0$

$x(x + 2)(x - 1) = 0$

Our result is a product of three factors: x, $(x + 2)$, and $(x - 1)$. Therefore, the equation has three solutions: 0, −2, and 1.

From this example, we can learn a general rule:

If you have an equation such as the following:
expression = 0
and if you can factor an x out of the expression,
then $x = 0$ is a solution of the equation.

Be careful not to just divide both sides by x!

This division improperly eliminates the solution $x = 0$. You are only allowed to divide by a variable if you are sure that the variable is NOT zero. (After all, you can't divide by zero, even in theory.)

Going in Reverse: Use FOIL

GMAT problems that involve quadratic equations often require you to work in reverse. That is, instead of starting with an equation and factoring it, you may need to start with factors and rewrite them as a quadratic equation. To do this, you need to use a relatively simple multiplication process called FOIL, an acronym for the four steps of the process: First, Outer, Inner, Last.

Reversing the process is generally an effective first step towards a solution. Factoring and distributing (using FOIL) are reverse processes.

To change the expression $(x + 7)(x - 3)$ into a quadratic equation, use FOIL as follows:

First: Multiply the first term of each factor together: $x \cdot x = x^2$

Outer: Multiply the outer terms of the expression together: $x(-3) = -3x$

Inner: Multiply the inner terms of the expression together: $7(x) = 7x$

Last: Multiply the last term of each factor together: $7(-3) = -21$

Now, there are 4 terms: $x^2 - 3x + 7x - 21$. By combining the two middle terms, we have our quadratic expression: $x^2 + 4x - 21$.

If a GMAT problem is written in the form of a quadratic equation, try factoring it. On the other hand, if a GMAT problem is written as the product of factors, try using FOIL to rewrite it as a quadratic equation.

Using FOIL with Square Roots

Some GMAT problems ask you to solve factored expressions that involve roots. For example, the GMAT might ask you to solve the following:

What is the value of $(\sqrt{8} - \sqrt{3})(\sqrt{8} + \sqrt{3})$?

Even though these problems don't involve any variables, you can solve them just like you would solve a pair of quadratic factors: use FOIL.

FIRST: $\sqrt{8} \cdot \sqrt{8} = 8$

OUTER: $\sqrt{8} \cdot \sqrt{3} = \sqrt{24}$

INNER: $\sqrt{8} \cdot \left(-\sqrt{3}\right) = -\sqrt{24}$

LAST: $\left(-\sqrt{3}\right)\left(\sqrt{3}\right) = -3$

The 4 terms are: $8 + \sqrt{24} - \sqrt{24} - 3$.

We can simplify this expression by removing the two middle terms (they cancel each other out) and subtracting: $8 + \sqrt{24} - \sqrt{24} - 3 = 8 - 3 = 5$.

Although the problem looks complex, using FOIL reduces the entire expression to 5.

1-Solution Quadratics

Not all quadratic equations have 2 solutions. Some have only 1 solution. Consider the following examples:

$x^2 + 8x + 16 = 0$ Here, the one solution for x is -4.
$(x + 4)(x + 4) = 0$

$x^2 - 6x + 9 = 0$ Here, the one solution for x is 3.
$(x - 3)(x - 3) = 0$

$x^2 + 2x + 1 = 0$ Here, the one solution for x is -1.
$(x + 1)(x + 1) = 0$

A fraction can NEVER have a denominator of zero.

Be careful not to assume that a quadratic equation always has two solutions. Always factor quadratic equations to determine their solutions. In doing so, you will see if a quadratic equation has 1 or 2 solutions.

Zero In the Denominator: Undefined

Math convention does not allow division by 0. When 0 appears in the denominator of an expression, then that expression is undefined. How does this convention affect quadratic equations? Consider the following:

What are the solutions to the following equation?

$$\frac{x^2 + x - 12}{x - 2} = 0$$

We notice a quadratic equation in the numerator. Since it is a good idea to start solving quadratic equations by factoring, we will factor this numerator as follows:

$$\frac{x^2 + x - 12}{x - 2} = 0 \rightarrow \frac{(x - 3)(x + 4)}{x - 2} = 0$$

If either of the factors in the numerator is 0, then the entire expression becomes 0. Thus, the solutions to this equation are $x = 3$ or $x = -4$.

Note that making the denominator of the fraction equal to 0 would NOT make the entire expression equal to 0. Recall that if 0 appears in the denominator, the expression becomes undefined. Thus, $x = 2$ (which would make the denominator equal to 0) is NOT a solution to this equation.

The 3 Favorite Forms

There are 3 quadratic-like expressions that come up so often on the GMAT that it pays to memorize them. You should immediately recognize these 3 expressions and know how to factor each one by heart.

Favorite #1: $x^2 - y^2 = (x + y)(x - y)$

Favorite #2: $x^2 + 2xy + y^2 = (x + y)(x + y)$

Favorite #3: $x^2 - 2xy + y^2 = (x - y)(x - y)$

When you see one of these forms, factor it immediately.

Whenever you see one of the three favorites in a GMAT problem, your first instinct should be to factor it! In general, this will put you on the path towards the solution to the problem.

Simplify: $\frac{x^2 + 4x + 4}{x^2 - 4}$, given that x does not equal 2 or -2.

Both the numerator and denominator of this fraction can be factored:

$$\frac{(x + 2)(x + 2)}{(x + 2)(x - 2)}$$

The numerator of the fraction is Favorite Form #2.
The denominator of the fraction is Favorite Form #1.

The expression $x + 2$ can be cancelled out from the numerator and denominator:

$$\frac{x^2 + 4x + 4}{x^2 - 4} = \frac{x + 2}{x - 2}$$

Problem Set

Solve the following problems. Use FOIL and factoring when needed.

1. If -4 is a solution for x in the equation $x^2 + kx + 8 = 0$, what is k?

2. Given that $\frac{d}{4} + \frac{8}{d} + 3 = 0$, what is d?

3. If 8 and -4 are the solutions for x, which of the following could be the equation?
(A) $x^2 - 4x - 32 = 0$ (B) $x^2 - 4x + 32 = 0$ (C) $x^2 + 4x - 12 = 0$
(D) $x^2 + 4x + 32 = 0$ (E) $x^2 + 4x + 12 = 0$

4. Given that $\frac{x^2 + 6x + 9}{x + 3} = 7$, what is x?

5. Given that $16 - y^2 = 10(4 + y)$, what is y?

6. Given that $x^2 - 10 = -1$, what is x?

7. Given that $\frac{(x + y)(x - y) + y^2}{8} = 2$, what is x?

8. Given that $t^2 + 2t + 1 = 18$, what is t?

9. If $x^2 + k = G$ and x is an integer, which of the following could be the value of $G - k$?
(A) 7 (B) 8 (C) 9 (D) 10 (E) 11

10. If the area of a certain square (expressed in square meters) is added to its perimeter (expressed in meters), the sum is 77. What is the length of a side of the square?

11. Hugo lies on top of a building, throwing pennies straight down to the street below. The formula for the height, H, that a penny falls is $H = Vt + 5t^2$, where V is the original velocity of the penny (how hard Hugo throws it when it leaves his hand) and t is equal to the time it takes to hit the ground. The building is 60 meters high, and Hugo throws the penny down at an initial speed of 20 meters per second. How long does it take for the penny to hit the ground?

12. $(3 - \sqrt{7})(3 + \sqrt{7}) =$

13. If $x^2 - 6x - 27 = 0$ and $y^2 - 6y - 40 = 0$, what is the maximum value of $x + y$?

14. Given that $\frac{x^3 + 4x^2 + 4x}{x} = 9$, what is x?

15. Given that $x^2 - 10x + 25 = 16$, what is x?

1. **6:** If -4 is a solution, then we know that $(x + 4)$ must be one of the factors of the quadratic equation. The other factor is $(x + ?)$. We know that the product of 4 and ? must be equal to 8; thus, the other factor is $(x + 2)$. We know that the sum of 4 and 2 must be equal to k. Therefore, $k = 6$.

2. $\boldsymbol{d = \{-8, -4\}}$**:** Multiply the entire equation by $4d$ (to eliminate the denominators) and factor.

$$d^2 + 32 + 12d = 0$$
$$d^2 + 12d + 32 = 0$$
$$(d + 8)(d + 4) = 0$$
$$d + 8 = 0 \quad \text{OR} \quad d + 4 = 0$$
$$d = -8 \qquad\qquad d = -4$$

3. **(A):** If the solutions to the equation are 8 and -4, the factored form of the equation is: $(x - 8)(x + 4)$.

Use FOIL to find the quadratic form: $x^2 - 4x - 32$. Therefore, the correct equation is (A).

4. $\boldsymbol{x = 4}$**:** Cross-multiply, simplify, and factor to solve.

$$\frac{x^2 + 6x + 9}{x + 3} = 7$$

$$x^2 + 6x + 9 = 7x + 21$$
$$x^2 - x - 12 = 0$$
$$(x + 3)(x - 4) = 0$$
$$x + 3 = 0 \quad \text{OR} \quad x - 4 = 0$$
$$x = -3 \qquad\qquad x = 4$$

Discard -3 as a value for x, since this value would make the denominator zero; thus, the fraction would be undefined.

5. $\boldsymbol{y = \{-4, -6\}}$**:** Simplify and factor to solve.

$$16 - y^2 = 10(4 + y)$$
$$16 - y^2 = 40 + 10y$$
$$y^2 + 10y + 24 = 0$$
$$(y + 4)(y + 6) = 0$$
$$y + 4 = 0 \quad \text{OR} \quad y + 6 = 0$$
$$y = -4 \qquad\qquad y - -6$$

6. $\boldsymbol{x = \{-3, 3\}}$**:**

$$x^2 - 10 = -1$$
$$x^2 = 9$$
$$x = \{-3, 3\}$$

7. $x = \{-4, 4\}$:

$$\frac{(x+y)(x-y)+y^2}{8} = 2$$

$$(x+y)(x-y)+y^2 = 16$$
$$x^2 - y^2 + y^2 = 16$$
$$x^2 = 16$$
$$x = \{-4, 4\}$$

8. $t = \{3\sqrt{2} - 1, -3\sqrt{2} - 1\}$:

$$t^2 + 2t + 1 = 18$$
$$(t+1)(t+1) = 18$$
$t + 1 = \sqrt{18}$ OR $t + 1 = -\sqrt{18}$
$t = \sqrt{18} - 1$ OR $t = -\sqrt{18} - 1$
$t = 3\sqrt{2} - 1$ OR $t = -3\sqrt{2} - 1$

In this case, since we recognize a common quadratic form on the left side of the equation ($x^2 + 2xy + y^2$, where $x = t$ and $y = 1$), we will factor the left side of the equation first, instead of trying to set everything equal to zero.

9. **(C):**

If $x^2 + k = G$
$x^2 = G - k$

In order for x to be an integer, $G - k$ must be a perfect square. The only perfect square among the answer choices is (C) 9.

10. $s = 7$: The area of the square = s^2. The perimeter of the square = $4s$.

$$s^2 + 4s = 77$$
$$s^2 + 4s - 77 = 0$$
$$(s + 11)(s - 7) = 0$$
$s + 11 = 0$ OR $s - 7 = 0$
$s = -11$ $s = 7$

Since the edge of a square must be positive, discard the negative value for s.

11. $t = 2$:

$$H = Vt + 5t^2$$
$$60 = 20t + 5t^2$$
$$5t^2 + 20t - 60 = 0$$
$$5(t^2 + 4t - 12) = 0$$
$$5(t + 6)(t - 2) = 0$$
$t + 6 = 0$ OR $t - 2 = 0$
$t = -6$ $t = 2$

Since a time must be positive, discard the negative value for t.

12. **2:**

Use FOIL to simplify this product:

F: $3 \times 3 = 9$

O: $3 \times \sqrt{7} = 3\sqrt{7}$

I: $-\sqrt{7} \times 3 = -3\sqrt{7}$

L: $-\sqrt{7} \times \sqrt{7} = -7$

$9 + 3\sqrt{7} - 3\sqrt{7} - 7 = 2$

Alternatively, recognize that the original expression is in the form $(x - y)(x + y)$, which is one of the three favorite forms and which equals $x^2 - y^2$ (the difference of two squares). Thus, the expression simplifies to $3^2 - (\sqrt{7})^2 = 9 - 7 = 2$.

13. **19:** Factor both quadratic equations. Then, use the largest possible values of x and y to find the maximum value of the sum $x + y$.

$$x^2 - 6x - 27 = 0$$
$$(x + 3)(x - 9) = 0$$
$$x + 3 = 0 \quad \text{OR} \quad x - 9 = 0$$
$$x = -3 \qquad\qquad x = 9$$

$$y^2 - 6y - 40 = 0$$
$$(y + 4)(y - 10) = 0$$
$$y + 4 = 0 \quad \text{OR} \quad y - 10 = 0$$
$$y = -4 \qquad\qquad y = 10$$

The maximum possible value of $x + y = 9 + 10 = 19$.

14. **$x = \{-5, 1\}$:**

$$\frac{x^3 + 4x^2 + 4x}{x} = 9$$

$$x^3 + 4x^2 + 4x = 9x$$
$$x^3 + 4x^2 - 5x = 0$$
$$x(x^2 + 4x - 5) = 0$$
$$x(x + 5)(x - 1) = 0$$
$$x = 0 \quad \text{OR} \quad x + 5 = 0 \quad \text{OR} \quad x - 1 = 0$$
$$x = -5 \qquad\qquad x = 1$$

However, the solution $x = 0$ renders the fraction in the original equation undefined, as we know the denominator of the fraction cannot be 0. Therefore, $x = \{-5, 1\}$.

15. **$x = \{1, 9\}$:**

$$x^2 - 10x + 25 = 16$$
$$x^2 - 10x + 9 = 0$$
$$(x - 9)(x - 1) = 0$$
$$x - 9 = 0 \quad \text{OR} \quad x - 1 = 0$$
$$x = 9 \qquad\qquad x = 1$$

Chapter 4
of
EQUATIONS, INEQUALITIES, & VIC's

FORMULAS

In This Chapter . . .

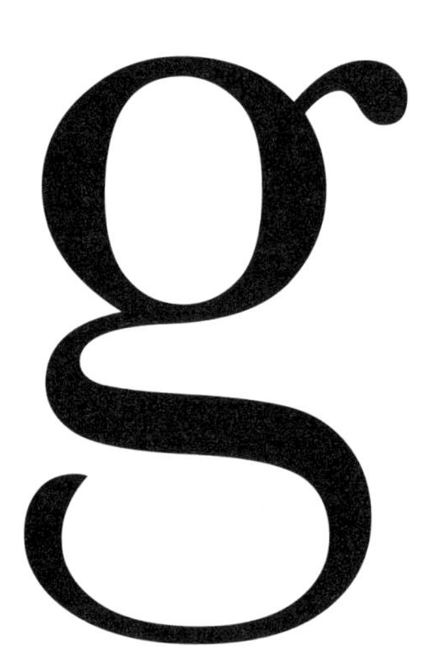

- Plug-In Formulas
- Strange Symbol Formulas
- Formulas with Unspecified Amounts
- Sequence Formulas
- Defining Rules for Sequences
- Sequence Problems: Alternate Method
- Sequences and Patterns

FORMULAS

Formulas are another means by which the GMAT tests your ability to work with unknowns. Formulas are specific equations that can involve multiple variables. There are 4 major types of Formula problems which the GMAT tests:

(1) Plug-in Formulas
(2) Strange Symbol Formulas
(3) Formulas with Unspecified Amounts
(4) Sequence Formulas

The GMAT uses formulas both in abstract problems and in real-life word problems. Becoming adept at working with formulas of all kinds is critical to your GMAT success.

Plug-in formulas are really a matter of careful computation.

Plug-In Formulas

The most basic GMAT formula problems provide you with a formula and ask you to solve for one of the variables in the formula by plugging in given values for the other variables. For example:

> **The formula for determining an individual's comedic aptitude, C, on a given day is defined as $QL \div J$, where J represents the number of jokes told, Q represents the overall joke quality on a scale of 1 to 10, and L represents the number of individual laughs generated. If Bella told 12 jokes, generated 18 laughs, and earned a comedic aptitude of 10.5, what was the overall quality of her jokes?**

Solving this problem simply involves plugging the given values into the formula in order to solve for the unknown variable Q:

$$C = \frac{QL}{J} \rightarrow 10.5 = \frac{18Q}{12} \rightarrow 10.5(12) = 18Q \rightarrow Q = \frac{10.5(12)}{18} \rightarrow Q = 7$$

The quality of Bella's jokes was rated a 7.

Strange Symbol Formulas

Another type of GMAT formula problem involves the use of strange symbols. In these problems, the GMAT introduces some arbitrary symbol, which defines a certain procedure. These problems may look confusing because of the unfamiliar symbols. However, the symbol is irrelevant. All that is important is that you carefully follow each step in the procedure which the symbol signals. Consider the following example:

$W \psi F = \left(\sqrt{F}\right)^{W}$ for all integers W and F. What is $4 \psi 9$?

The symbol ψ between two numbers signals the following procedure: take the square root of the second number and then raise that value to the power of the first number.

Don't be fooled by strange symbols. You will usually be told what they mean.

Thus, $4 \psi 9 = \left(\sqrt{9}\right)^4 = 3^4 = 81$.

More challenging strange-symbol problems require you to use the given procedure more than once. For example:

What is $4 \psi (3 \psi 16)$?

Always perform the procedure inside the parentheses FIRST:

$3 \psi 16 = \left(\sqrt{16}\right)^3 = 4^3 = 64$.

Now we can rewrite the original formula as follows: $4 \psi (3 \psi 16) = 4 \psi 64$.

Performing the procedure a second time yields the answer:

$4 \psi 64 = \left(\sqrt{64}\right)^4 = 8^4 = 4{,}096$.

Formulas with Unspecified Amounts

The most challenging formula problems on the GMAT are those that involve unspecified amounts. Typically, these questions focus on the increase or decrease in the value of a certain formula, given a change in the value of the variables. Just as with other GMAT problems with unspecified amounts, solve these problems by PICKING NUMBERS!

If the length of the side of a cube decreases by two-thirds, by what percentage will the volume of the cube decrease?

First, consider the formula involved here. The volume of a cube is defined by the formula $V = s^3$, where s represents the length of a side. Then, pick a number for the length of the side of the cube.

Formula problems that involve unspecified amounts should be solved by picking smart numbers for the unspecified variables.

Let's say the cube has a side of 9 units. Note that this is a "smart" number to pick because it is divisible by 3 (the denominator of two-thirds).

Then, its volume $= s^3 = 9 \times 9 \times 9 = 729$.

If the cube's side decreases by two-thirds, its new length is $9 - \frac{2}{3}(9) = 9 - 6 = 3$ units.

Its new volume $= s^3 = 3 \times 3 \times 3 = 27$.

We determine percentage decrease as follows:

$$\frac{\text{change}}{\text{original}} = \frac{729 - 27}{729} = \frac{702}{729} \approx .963 = 96.3\% \text{ decrease.}$$

Sequence Formulas

The final type of GMAT formula problem involves sequences. A sequence is a collection of numbers in a set order. The order of a given sequence is determined by a RULE. Here are examples of sequence RULES:

$A_n = 9n$ The nth term of this sequence is defined by the rule $9n$. For example, the third term in this sequence is $9n = 9(3) = 27$. The fourth term in this sequence is $9n = 9(4) = 36$, etc. The first ten terms of the sequence are as follows:

9, 18, 27, 36, 45, 54, 63, 72, 81, 90

Sequences are defined by function rules, where each term is a function of its place in the sequence.

$S_n = n + 3$ The nth term of this sequence is defined by the rule $n + 3$. For example, the first term in this sequence is $1 + 3 = 4$. The second term in this sequence is $2 + 3 = 5$. The first ten terms of the sequence are as follows:

4, 5, 6, 7, 8, 9, 10, 11, 12, 13

$Q_n = n^2 + 4$ The nth term of this sequence is defined by the rule $n^2 + 4$. For example, the first term in this sequence is $1^2 + 4 = 5$. The second term in this sequence is $2^2 + 4 = 8$. The first ten terms of the sequence are as follows:

5, 8, 13, 20, 29, 40, 53, 68, 85, 104

Defining Rules for Sequences

The important thing to remember about sequence problems is that YOU MUST be given the RULE in order to find a particular number in a sequence. It is tempting (BUT WRONG!) to try to solve sequence problems without the RULE. For example:

If $S_7 = 5$ and $S_8 = 6$, what is S_9?

It looks like this is a sequence that counts by 1's, so it is tempting to say that $S_9 = 7$. However, this deduction is NOT mathematically valid, because we were not given the RULE for this particular sequence. The sequence might be 5, 6, 7, 8, 9... However, it might also be 5, 6, 8, 11, 15. Without the rule, there is no way to be sure.

Given (1) information about the form of the sequence and (2) at least one term, you can find the rule using some trial and error. However, there are some guidelines to follow:

(1) If you are told that the difference between successive terms is always the same, the rule will take the form of $kn + x$, where k and x are real numbers and k is equal to the difference between successive terms.

(2) If you are told that the difference between *the difference* between successive terms is always the same, the rule will take the form of $an^2 + bn + c$, where a, b, and c are real numbers.

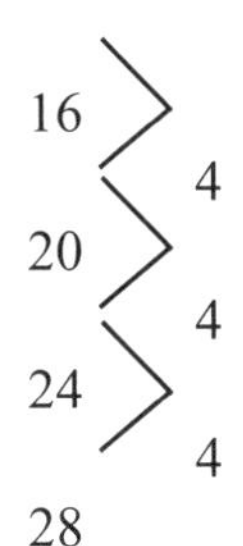

The difference between successive terms is the same. The rule for this sequence is $kn + x$, where $k = 4$.

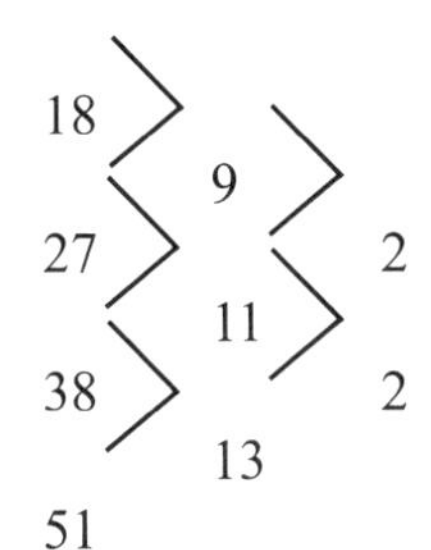

The difference between *the difference* between successive terms is the same. The rule for this sequence is $an^2 + bn + c$.

Consider the two sequences above.

In the first sequence, assume that we are told that the difference between each successive term is the constant. This means that the equation for this sequence must be in the form $4n + x$. We can find the value of x by using any of the terms in the sequence. For example:

Many sequences are of the form $kn + x$, where k equals the difference between successive terms.

The first term in the sequence ($n = 1$) has a value of 16.
Set $4(1) + x = 16$
$x = 12$

We can confirm that $x = 12$ by using another term in the sequence.
For example, the second term in the sequence ($n = 2$) has a value of 20. This verifies that $x = 12$, since $4(2) + 12 = 20$.

Therefore, $4n + 12$ is the rule for the first sequence above. We can use this formula to find any term in the sequence.

The second sequence is a bit more complicated. Assume that we are told that the difference between the *difference* of each successive term is constant. Since the difference between the *difference* between successive terms is the same, we can set up a system of equations for this sequence using the formula $an^2 + bn + c$ as follows:

The first term in the sequence ($n = 1$) has a value of 18: $a(1)^2 + b(1) + c = 18$.
The second term in the sequence ($n = 2$) has a value of 27: $a(2)^2 + b(2) + c = 27$.
The third term in the sequence ($n = 3$) has a value of 38: $a(3)^2 + b(3) + c = 38$.

We have three different equations with three unknowns. Using substitution or another method, we can solve this system of equations for the three unknowns: $a = 1$, $b = 6$, and $c = 11$. Therefore, the formula for this sequence is $n^2 + 6n + 11$. We can use this formula to find any term in the sequence.

It is important to remember that you can derive a rule for a sequence only if you are provided with BOTH (1) information about the form of the sequence (e.g. "the difference between each successive term in the sequence is constant" or "the sequence takes the form $kn + x$") and (2) some consecutive terms in the sequence.

Sequence Problems: Alternate Method

For simple additive sequences, in which the same number is added to each term to yield the next term, you can use the following alternative method:

If each number in a sequence is three more than the previous number, and the sixth number is 32, what is the 100th number?

Instead of finding the rule for this sequence, consider the following reasoning:

From the sixth to the one hundredth term, there are 94 "jumps" of 3. Since $94 \times 3 = 282$, there is an increase of 282 from the sixth term to the one hundredth term:

If a problem seems to require too much computation, look for an easier way to solve it.

$32 + 282 = 314$.

Sequences and Patterns

Some sequences are easier to look at in terms of patterns, rather than rules. For example, consider the following:

If $S_n = 3^n$, what is the units digit of S_{65}?

Clearly, you cannot be expected to multiply out 3^{65} on the GMAT. Therefore, you must assume that there is a pattern in the powers of three.

$3^1 = \mathbf{3}$
$3^2 = \mathbf{9}$
$3^3 = 2\mathbf{7}$
$3^4 = 8\mathbf{1}$
$3^5 = 24\mathbf{3}$
$3^6 = 72\mathbf{9}$
$3^7 = 2{,}18\mathbf{7}$
$3^8 = 6{,}56\mathbf{1}$

Note the pattern of the units digits in the powers of 3: 1, 3, 9, 7, [repeating]... Also note that the units digit of S_n, when n is a multiple of 4, is always equal to 1. You can use the multiples of 4 as "anchor points" in the pattern. Since 65 is 1 more than 64 (the closest multiple of 4), the units digit of S_{65} will be 3, which always follows 1 in the pattern.

Problem Set

1. Given that $A * B = 4A - B$, what is the value of $(3 * 2) * 3$?

2. Given that $u \times y$ (with x on top and z on bottom) $= \frac{u + y}{x + z}$, what is 8×10 (with 4 on top and 5 on bottom)?

3. Given that $\boxed{\begin{matrix} A \\ B \end{matrix}} = A^2 + B^2 + 2AB$, what is $A + B$, if $\boxed{\begin{matrix} A \\ B \end{matrix}} = 9$?

For problem #4, use the following information: $x \Longrightarrow y = x + (x + 1) + (x + 2) \ldots + y$.
e.g. $3 \Longrightarrow 7 = 3 + 4 + 5 + 6 + 7$.

4. What is the value of $(100 \Longrightarrow 150) - (125 \Longrightarrow 150)$?

5. Life expectancy is defined by the formula $\frac{2SB}{G}$, where S = shoe size, B = average monthly electric bill in dollars, and G = GMAT score. If Elvin's GMAT score is twice his monthly electric bill, and his life expectancy is 50, what is his shoe size?

6. The formula for spring factor in a shoe insole is $\frac{w^2 + x}{3}$, where w is the width of the insole in centimeters and x is the grade of rubber on a scale of 1 to 9. What is the maximum spring factor for an insole that is 3 centimeters wide?

7. Cost is expressed by the formula tb^4. If b is doubled, by what factor has the cost increased?
(A) 2 (B) 6 (C) 8 (D) 16 (E) ½

8. If the scale model of a cube sculpture is .5 cm per every 1 m of the real sculpture, what is the volume of the model, if the volume of the real sculpture is 64 m^3?

For problems #9-10, use the following information: The "competitive edge" of a baseball team is defined by the formula $\frac{W^3}{L^2}$, where W represents the number of wins, and L represents the number of losses.

9. This year, the GMAT All-Stars tripled the number of wins and halved the number of losses that they had last year. By what factor did their "competitive edge" increase?

10. The "competitive edge" of the LSAT Juveniles was 256 times as high in year A as it was in year B. If the team won one quarter as many games in year B as it did in year A, what was the percentage increase from year A to year B in the number of games it lost?

11. If the radius of a circle is tripled, what is the ratio of the area of half the original circle to the area of the whole new circle?

For problems #12-13, use the following sequence: $A_n = 3 - 8n$.

12. What is A_1?

13. What is $A_{11} - A_9$?

14. If each number in a sequence is three more than the previous number, and the eighth number is 46, what is the rule for this sequence?

15. If $S_n = 4^n + 5^{n+1} + 3$, what is the units digit of S_{100}?

1. **37:** First, simplify 3 * 2: 4(3) – 2 = 12 – 2 = 10. Then, solve 10 * 3: 4(10) – 3 = 40 – 3 = 37.

2. **2:** Plug the numbers in the grid into the formula, matching up the number in each section with the corresponding variable in the formula $\frac{u+y}{x+z} = \frac{8+10}{4+5} = \frac{18}{9} = 2$.

3. ***A* + *B* = {3, –3}:**

$$A^2 + B^2 + 2AB = 9$$
$$(A + B)^2 = 9$$
$$A + B = 3 \quad \text{OR} \quad A + B = -3$$

First, set the formula equal to 9. Then, factor the expression $A^2 + B^2 + 2AB$. Unsquare both sides, taking both the positive and negative roots into account.

4. **2,800:** This problem contains two components: the sum of all the numbers from 100 to 150, and the sum of all the numbers from 125 to 150. Since we are finding the difference between these two components, we are essentially finding just the sum of all the numbers from 100 to 124. You can think of this logically by solving a simpler problem: find the difference between (1 ==> 5) – (3 ==> 5).

$$\begin{array}{rr} & 1+2+3+4+5 \\ - & 3+4+5 \\ \hline & 1+2 \quad\quad\quad\quad \end{array}$$

There are 25 numbers from 100 to 124 (124 – 100 + 1). Remember to add one before you're done! To find the sum of these numbers, multiply by the average term: (100 + 124) ÷ 2 = 112. 25 × 112 = 2,800.

5. **Size 50:**

$$\frac{2SB}{2B} = 50$$
$$S = 50$$

Substitute $2B$ for G in the formula. Note that the term $2B$ appears in both the numerator and denominator, so they cancel out.

6. **6:** Determine the maximum spring factor by setting $x = 9$.
Let s = spring factor

$$s = \frac{w^2 + x}{3}$$

$$s = \frac{(3)^2 + 9}{3} = \frac{18}{3} = 6$$

7. **(D):** Pick numbers to see what happens to the cost when b is doubled. If the original value of b is 2, the cost is $16t$. When b is doubled to 4, the new cost value is $256t$. The cost has increased by a factor of $\frac{256}{16}$, or 16.

8. **8 cm³**:

$V = s^3$
$64 = s^3$
$s = 4$ The length of a side on the real sculpture is 4 m.

$\frac{.5 \text{ cm}}{1 \text{ m}} = \frac{x \text{ cm}}{4 \text{ m}}$

$x = 2$ The length of a side on the model is 2 cm.

$V = s^3 = (2)^3 = 8$ The volume of the model is 8.

9. **108:**

Let c = competitive edge

$c = \frac{W^3}{L^2}$

Pick numbers to see what happens to the competitive edge when W is tripled and L is halved. If the original value of W is 2 and the original value of L is 4, the original value of c is .5. If you triple W to 6 and halve L to 2, the new value of c is 54. The competitive edge has increased from .5 to 54.

$.5x = 54 \rightarrow x = 108$ The competitive edge has increased by a factor of 108.

10. **100% increase:**

Let A = the competitive edge in year A
Let B = the competitive edge in year B

Assign variables to the competitive edge in both year A and year B. Then, express the information given in the problem in algebraic equations.

$W_A = 4W_B$
$A = 256B$

Substitute $4W_B$ for W_A.

$\frac{(4W_B)^3}{L_A^2} = \frac{256W_B^3}{L_B^2}$

Cross-multiply and solve. Don't forget to raise both the number and the variable to the third power.

$64W_B^3L_B^2 = 256W_B^3L_A^2$
$L_B^2 = 4L_A^2$
$L_B = 2L_A$

Unsquare both sides. Unsquare the number and the variable!

The number of losses in year B is double the number of losses in year A. Therefore, the percent increase is 100%.

11. $\frac{1}{18}$**:** Pick real numbers to solve this problem. Set the radius of the original circle equal to 2. Therefore, the radius of the new circle is equal to 6. Once you compute the areas of both circles, you can find the ratio:

Original Circle	New Circle
$r = 2$ $A = \pi r^2$ $= 4\pi$	$r = 6$ $A = \pi r^2$ $= 36\pi$

$$\frac{\text{Area of half the original circle}}{\text{Area of the new circle}} = \frac{2\pi}{36\pi} = \frac{1}{18}$$

12. **–5:** $A_n = 3 - 8n$
$A_1 = 3 - 8(1) = 3 - 8 = -5$

13. **–16:** $A_n = 3 - 8n$
$A_{11} = 3 - 8(11) = 3 - 88 = -85$
$A_9 = 3 - 8(9) = 3 - 72 = -69$
$A_{11} - A_9 = -85 - (-69) = -16$

14. **3*n* + 22:** Set up a partial sequence table in order to derive the rule for this sequence. Because the difference between successive terms in the sequence is always 3, you know that the rule will be of the form $3n + x$. You can then solve for x using any term in the sequence. Using the fact that the 8th term has a value of 46, we know that $3(8) + x = 46$. Thus, $x = 22$ and the rule for this sequence is $3n + 22$.

n	value
6	40
7	43
8	46
9	49

15. **4:** Begin by listing the first few terms in the sequence, so that you can find a pattern:

$S_1 = 4^1 + 5^{1+1} + 3 = 4 + 25 + 3 = 3\mathbf{2}$
$S_2 = 4^2 + 5^{2+1} + 3 = 16 + 125 + 3 = 14\mathbf{4}$
$S_3 = 4^3 + 5^{3+1} + 3 = 64 + 625 + 3 = 69\mathbf{2}$
$S_4 = 4^4 + 5^{4+1} + 3 = 256 + 3{,}125 + 3 = 3{,}38\mathbf{4}$

The units digit of all odd-numbered terms is 2. The units digit of all even-numbered terms is 4. Since S_{100} is an even-numbered term, its units digit will be 4.

Chapter 5

of

EQUATIONS, INEQUALITIES, & VIC's

FUNCTIONS

In This Chapter . . .

- Numerical Substitution
- Variable Substitution
- Compound Functions
- Functions with Unknown Constants
- Function Graphs

FUNCTIONS

Functions are very much like the "magic boxes" you may have learned about in elementary school.

You put a 2 into the magic box, and a 7 comes out. You put a 3 into the magic box, and a 9 comes out. You put a 4 into the magic box, and an 11 comes out. What is the magic box doing to your number?

There are many possible ways to describe what the magic box is doing to your number. One possibility is as follows: The magic box is doubling your number and adding 3.

$2(2) + 3 = 7$
$2(3) + 3 = 9$
$2(4) + 3 = 11$

A function rule describes a series of operations to be performed on a variable.

This description would yield the following "rule" for this magic box: $2x + 3$. This rule can be written in function form as:

$f(x) = 2x + 3$.

Note that this rule may or may not be the "true" rule for the magic box. That is, if we put more numbers into the box and watch what numbers emerge, this rule may or may not hold. It is never possible to generalize a rule using specific cases.

Nevertheless, the magic box analogy is a helpful way to conceptualize a function as a RULE built on an independent variable. The value of a function changes as the value of the variable changes. In other words, the value of a function is dependent on the value of the variable. Examples of functions include:

$f(x) = 4x^2 - 11$ The value of the function, f, is dependent on the independent variable, x.

$g(t) = t^3 + \sqrt{t} - \frac{2t}{5}$ The value of the function, g, is dependent on the independent variable, t.

We can think of functions as consisting of an "input" variable (the number you put into the magic box), and a corresponding "output" value (the number that comes out of the magic box).

There are five types of function problems you may find on the GMAT:

(1) Numerical Substitution
(2) Variable Substitution
(3) Compound Functions
(4) Functions with Unknown Constants
(5) Function Graphs

Numerical Substitution

This is the most basic type of function problem. Input the numerical value (5) in place of the independent variable (x) to determine the value of the function.

If $f(x) = x^2 - 2$, what is the value of $f(5)$?

In this problem, you are given a rule for $f(x)$: square x and subtract 2. Then, you are asked to apply this rule to the number 5. Square 5 and subtract 2 from the result:

$$f(5) = (5)^2 - 2 = 25 - 2 = 23$$

When substituting a variable expression into a function, keep the expression inside parentheses.

Variable Substitution

This type of problem is slightly more complicated. Instead of finding the output value for a numerical input, you must find the output when the input is an algebraic expression.

If $f(g) = g^2 - \frac{g}{3}$, what is the value of $f(w + 6)$?

Input the variable expression ($w + 6$) in place of the independent variable (g) to determine the value of the function:

$$f(w+6) = (w+6)^2 - \frac{w+6}{3}$$

Compare this equation to the equation for $f(g)$. The expression ($w + 6$) has taken the place of every g in the original equation. In a sense, you are treating the expression ($w + 6$) as if it were a single letter or variable.

The rest is algebraic simplification:

$$f(w+6) = (w+6)(w+6) - \left(\frac{w}{3} + \frac{6}{3}\right)$$

$$= w^2 + 12w + 36 - \frac{w}{3} - 2$$

$$= w^2 + 11\frac{2}{3}w + 34$$

Compound Functions

Imagine putting a number into one magic box, and then putting the output directly into another magic box. This is the situation you have with compound functions.

If $f(x) = x^3 + \sqrt{x}$ and $g(x) = 4x - 3$, what is $f(g(3))$?

The key to solving compound function problems is taking them step by step. Work from the INSIDE OUT.

$g(3) = 4(3) - 3 = 12 - 3 = 9$

Use this inner function solution as the new input variable for the outer function.

$f(9) = (9)^3 + \sqrt{9} = 729 + 3 = 732$

Solve compound functions from the INSIDE OUT.

Note that changing the order of the compound functions changes the answer.

If $f(x) = x^3 + \sqrt{x}$ and $g(x) = 4x - 3$, what is $g(f(3))$?

$f(3) = (3)^3 + \sqrt{3} = 27 + \sqrt{3}$

Use this inner function solution as the new input variable for the outer function.

$g(27 + \sqrt{3}) = 4(27 + \sqrt{3}) - 3 = 108 + 4\sqrt{3} - 3 = 105 + 4\sqrt{3}$

Thus, $g(f(3)) = 105 + 4\sqrt{3}$.

Functions with Unknown Constants

On the GMAT, you may be given a function with an unknown constant. You will also be given the value of the function for a specific number. You can combine these pieces of information to find the complete function rule.

If $f(x) = ax^2 - x$, and $f(4) = 28$, what is $f(-2)$?

Solve these problems in three steps. FIRST, use the value of the input variable and the corresponding output value of the function to solve for the unknown constant:

The key to identifying function graphs is plotting points.

$$f(4) = a(4)^2 - 4 = 28$$
$$16a - 4 = 28$$
$$16a = 32$$
$$a = 2$$

THEN, rewrite the function, replacing the constant with its numerical value:

$$f(x) = ax^2 - x = 2x^2 - x$$

FINALLY, solve the function for the new input variable:

$$f(-2) = 2(-2)^2 - (-2) = 8 + 2 = 10$$

Function Graphs

A function can be visualized by graphing it in the coordinate plane. The input variable is considered the domain of the function, or the x-coordinate. The corresponding output is considered the range of the function, or the y-coordinate.

What is the graph of the function $f(x) = -2x^2 + 1$?

Create an INPUT-OUTPUT table by evaluating the function for several input values:

INPUT	OUTPUT	(x, y)
-3	$-2(-3)^2 + 1 = -17$	$(-3, -17)$
-2	$-2(-2)^2 + 1 = -7$	$(-2, -7)$
-1	$-2(-1)^2 + 1 = -1$	$(-1, -1)$
0	$-2(0)^2 + 1 = 1$	$(0, 1)$
1	$-2(1)^2 + 1 = -1$	$(1, -1)$
2	$-2(2)^2 + 1 = -7$	$(2, -7)$
3	$-2(3)^2 + 1 = -17$	$(3, -17)$

Then, plot the points to see the shape of the graph:

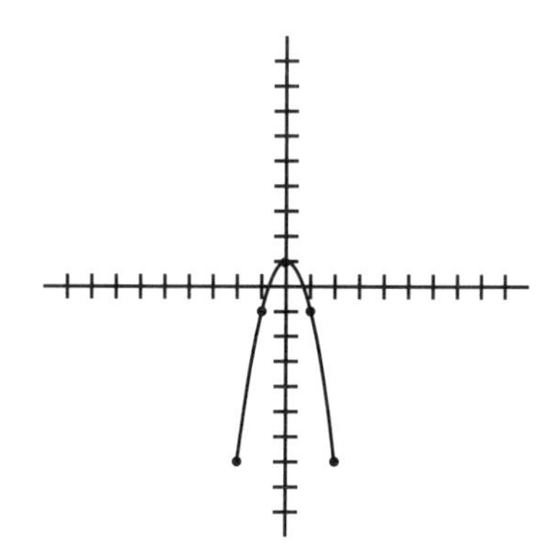

Problem Set

1. If $f(x) = 2x^4 - x^2$, what is the value of $f\left(2\sqrt{3}\right)$?

2. If $g(x) = 3x + \sqrt{x}$, what is the value of $g(d^2 + 6d + 9)$?

3. If $k(x) = 4x^3a$, and $k(3) = 27$, what is $k(2)$?

4. If $f(x) = 3x - \sqrt{x}$ and $g(x) = x^2$, what is $f(g(4))$?

5. If $f(x) = 3x - \sqrt{x}$ and $g(x) = x^2$, what is $g(f(4))$?

6. If $f(x) = 2x^2 - 4$ and $g(x) = 2x$, for what values of x will $f(x) = g(x)$?

7. If $f(x) = (x - 1)^2$, what is the value of $f(x - 1)$?
   (A) $x^2 + 2x + 1$ (B) $x^2 - 4x + 2$ (C) $x^2 - 4x + 4$ (D) $x^2 + 4x + 4$

8. If $f(x) = (x + \sqrt{3})^4$, what is the range of the function $f(x)$?
   (A) $\sqrt{3} < f(x) < 4$ (B) $f(x) \geq 0$ (C) $f(x) < 0$ (D) $f(x) \neq 0$

9. If $f(x) = 3x - 1$ and $g(x) = x^2$, what is $g(f(3x + 1))$?

10. If $g(x) = \dfrac{x^3 - ax}{4}$, and $g(2) = \dfrac{1}{2}$, what is the value of $g(4)$?

11. If $k(x) = x + 1$, and $j(x) = 2x - 4$, for what value will $k(x) = j(k(x))$?

For problems #12-15, match each function with its graph.

12. $f(x) = -x^2 + 5$

13. $g(x) = |x - 1| - 1$

14. $h(x) = \dfrac{x}{3} + 4$

15. $k(x) = x^3 + 1$

(A)

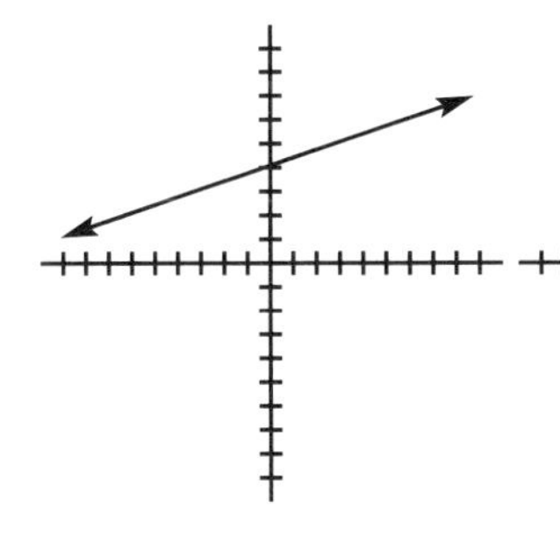

(B)

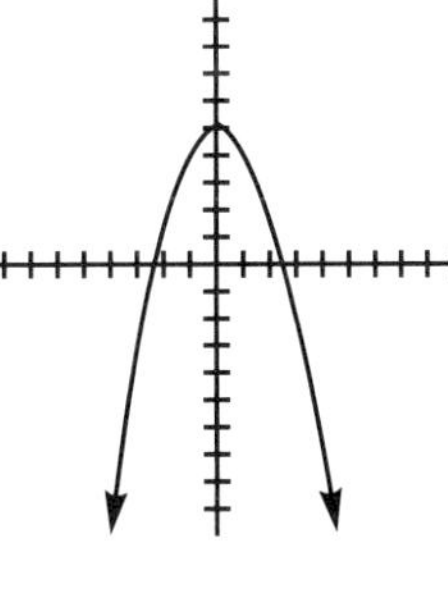

(C)

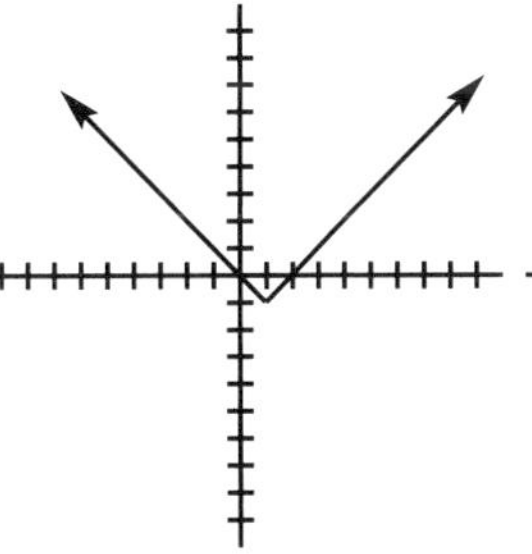

(D)

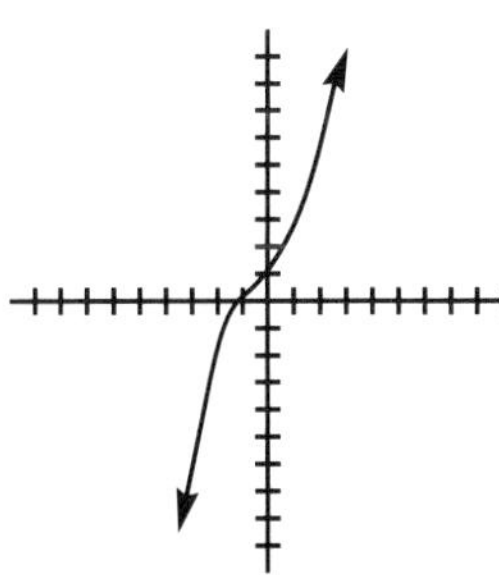

1. **276:** $f(x) = 2\left(2\sqrt{3}\right)^4 - \left(2\sqrt{3}\right)^2 = 2(2)^4\left(\sqrt{3}\right)^4 - (2)^2\left(\sqrt{3}\right)^2$
$= (2 \cdot 16 \cdot 9) - (4 \cdot 3)$
$= 288 - 12$
$= 276$

2. **$3d^2 + 19d + 30$:** $g(d^2 + 6d + 9) = 3(d^2 + 6d + 9) + \sqrt{d^2 + 6d + 9}$
$= 3d^2 + 18d + 27 + \sqrt{(d + 3)^2}$
$= 3d^2 + 18d + 27 + d + 3$
$= 3d^2 + 19d + 30$

3. **8:** $k(3) = 27$ Therefore,

$4(3)^3 a = 27 \rightarrow k(x) = 4x^3\left(\frac{1}{4}\right) = x^3 \rightarrow k(2) = (2)^3 = 8$
$108a = 27$
$a = \frac{1}{4}$

4. **44:** First, find the output value of the inner function: $g(4) = 16$.
Then, find $f(16)$: $3(16) - \sqrt{16} = 48 - 4 = 44$.

5. **100:** First, find the output value of the inner function: $f(4) = 3(4) - \sqrt{4} = 12 - 2 = 10$.
Then, find $g(10)$: $10^2 = 100$.

6. **$x = \{-1, 2\}$:** To find the values for which $f(x) = g(x)$, set the functions equal to each other.
$2x^2 - 4 = 2x$
$2x^2 - 2x - 4 = 0$
$2(x^2 - x - 4) = 0$
$2(x - 2)(x + 1) = 0$
$x - 2 = 0$ OR $x + 1 = 0$
$x = 2$ $x = -1$

7. **(C):** Plug the variable expression $x - 1$ into the rule for the function $f(x)$.
$f(x - 1) = [(x - 1) - 1]^2$
$= (x - 2)^2$
$= x^2 - 4x + 4$

Alternately, you might solve this problem as a VIC (see the VIC chapter in this book) by picking a number for x and evaluating $f(x)$ for $x - 1$. Then, you would evaluate each answer choice to find the one that matches your target number.

8. **(B):** If $f(x) = (x + \sqrt{3})^4$, the range of outputs, or y-values, can never be negative. Regardless of the value of x, raising $x + \sqrt{3}$ to an even power will result in a non-negative y-value. Therefore, the range of the function is all non-negative numbers, or $f(x) \geq 0$.

9. **$81x^2 + 36x + 4$:** First, find the output value of the inner function:

$$f(3x + 1) = 3(3x + 1) - 1$$
$$= 9x + 3 - 1 = 9x + 2$$

Then, find $g(9x + 2)$: $(9x + 2)^2 = (9x + 2)(9x + 2) = 81x^2 + 36x + 4$.

10. **13:** $g(2) = \frac{(2)^3 - a(2)}{4} = \frac{1}{2}$

$$8 - 2a = 2$$

$$2a = 6 \quad \rightarrow \quad g(x) = \frac{x^3 - 3x}{4} \quad \rightarrow \quad g(4) = \frac{(4)^3 - 3(4)}{4} = \frac{64 - 12}{4} = 13$$
$$a = 3$$

11. **$x = 3$:** First, simplify $j(k(x))$.

$$j(k(x)) = j(x + 1) = 2(x + 1) - 4 = 2x - 2$$

To find the values for which $k(x) = j(k(x))$, set the functions equal to each other.

$$x + 1 = 2x - 2$$
$$x = 3$$

12. **(B):**
$f(x) = -x^2 + 5$
This function is a parabola. You can identify the correct graph by plotting the y-intercept (0, 5).

13. **(C):**
$g(x) = |x - 1| - 1$
This function is a V-shape. You can identify the correct graph by plotting the y-intercept (0, 0).

14. **(A):**
$h(x) = \frac{x}{3} + 4$

This function is a straight line. You can identify the correct graph by plotting the y-intercept (0, 4).

15. **(D):**
$k(x) = x^3 + 1$
This function is a hyperbola. You can identify the correct graph by plotting the y-intercept (0, 1).

Chapter 6
of
EQUATIONS, INEQUALITIES, & VIC's

INEQUALITIES

In This Chapter . . .

- Just Like Equations, With One Exception
- Inequalities and Even Exponents
- Inequalities and Absolute Value
- Combining Inequalities: Line 'Em Up
- Plugging in the Extremes

INEQUALITIES

Unlike equations, which relate two equivalent quantities, inequalities compare quantities that have different values. Inequalities are used to express four kinds of relationships, illustrated by the following examples.

(1) x is less than 4

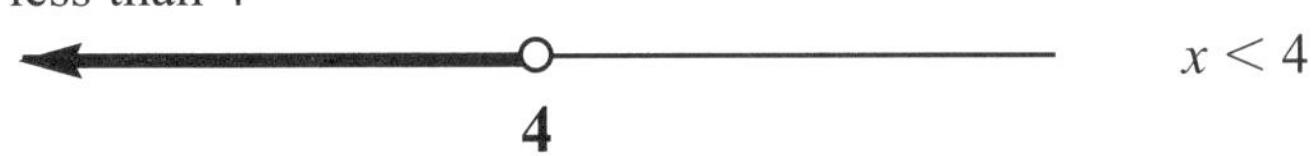

$x < 4$

(2) x is less than or equal to 4

$x \leq 4$

Note the difference between $x < 4$ and $x \leq 4$.

(3) x is greater than 4

$x > 4$

(4) x is greater than or equal to 4

$x \geq 4$

Number lines, like those shown above, are extremely useful tools when working with inequalities. They are an excellent way to visualize exactly what a given inequality means.

Just Like Equations, With One Exception

Inequalities can be solved just like equations: isolate the variable by performing identical operations to both sides of the inequality. There is, however, one procedure that makes inequalities different from equations: **When you multiply or divide an inequality by a negative number, the inequality sign flips!**

For example:

Given that $4 - 3x < 10$, what is the range of possible values for x?

$4 - 3x < 10$	In isolating x in this equation, we divide both sides by -3.
$-3x < 6$	Because we divide by a negative number, the inequality
$x > -2$	sign flips from **less than** to **greater than**!

Now consider the following example:

Given that $xy < 3y$, what is the range of possible values for x?

In order to isolate x in the above equation, it seems that we should simply divide both sides of the inequality by y. However, because we don't know whether y is positive or negative, we are NOT ALLOWED to divide both sides of the equation by y.

If y is positive, then the solution to the inequality is $x < 3$. However, if y is negative, we are dividing an inequality by a negative number; thus, the sign flips and yields the solution $x > 3$.

When you multiply or divide by a negative number or variable, you must flip the sign of the inequality.

This example leads us to the following rule: **Do not multiply or divide inequalities by unknowns (unless you know the sign of the unknown).** What you should do instead is **move all the terms to one side of the inequality and factor**:

$xy - 3y < 0$ → $y(x - 3) < 0$ → y and $x - 3$ have opposite signs

Inequalities and Even Exponents

Inequalities involving even exponents require you to consider TWO scenarios.

Consider the example $x^2 < 4$.

If x is positive, then $x < 2$. On the other hand, if x is negative, then $x > -2$. Notice that the inequality sign flips! Now, combine the two scenarios into one range of values as follows: $-2 < x < 2$. This states that x falls in between -2 and 2. On a number line, this is illustrated as follows:

Here is another example:

If $10 + x^2 \geq 91$, what is the range of possible values for x?

$$10 + x^2 \geq 91$$
$$x^2 \geq 81$$

If x is positive, then $x \geq 9$. If x is negative, then $x \leq -9$. Notice again that the inequality sign flips! This is because, in taking the square root of both sides, we divide the inequality by a negative number (x). One way of thinking about this is by using a number line. We can see that x can be any number except for those between -9 and 9.

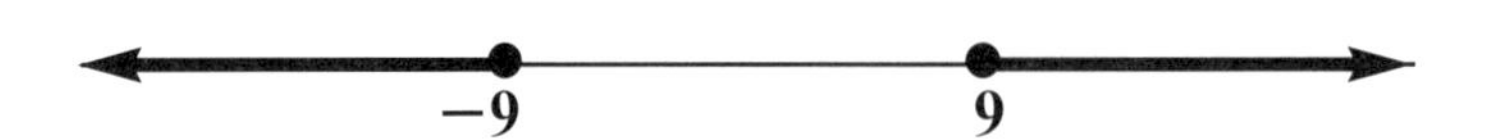

Inequalities and Absolute Value

Inequalities involving absolute value also require you to consider TWO scenarios. Just as with equations, inequalities with absolute value are solved by setting the terms inside the absolute value brackets to BOTH positive and negative values. For example:

Given that $|x - 5| < 8$, what is the range of possible values for x?

To work out the FIRST scenario, we simply remove the absolute value brackets and solve.

$$|x - 5| < 8$$
$$x - 5 < 8$$
$$x < 13$$

Solving absolute value inequalities yields a range of values.

To work out the SECOND scenario, we reverse the signs of the terms inside the absolute value brackets, remove the brackets, and solve again.

$$|x - 5| < 8$$
$$-(x - 5) < 8$$
$$-x + 5 < 8$$
$$-x < 3$$
$$x > -3$$

We can combine these two scenarios into one range of values for x (just as we did when thinking about the two scenarios for certain exponential inequalities). This range is illustrated by the following number line:

Note, by the way, that you would NEVER change $|x - 5|$ to $x + 5$. When you drop the absolute value signs, you either leave the expression alone OR enclose the ENTIRE expression in parentheses and put a negative sign in front.

Combining Inequalities: Line 'Em Up

Many GMAT inequality problems involve more than one inequality. To solve problems with multiple inequalities, solve and simplify the inequalities and line up the variables.

If $x > 8$, $x < 17$, and $x + 5 < 19$, what is the range of possible values for x?

First, solve any expressions that need to be solved. In this example, only the last inequality needs to be solved.

$x + 5 < 19$
$x < 14$

Second, simplify the inequalities so that all the inequality symbols point in the same direction, preferably to the left (less than).

$8 < x$
$x < 17$
$x < 14$

Third, line up the common variables in the inequalities

$8 < x$
$x < 17$
$x < 14$

Finally, combine the inequalities by taking the more limiting upper and lower extremes.

$8 < x < 14$

LICHLUC:
You can remember these steps with this acronym:
LI - List
CH - Change
LU - Line Up
C - Combine

Given that $u < t$, $b > r$, $f < t$, and $r > t$, is $b > u$?

Combine the 4 given inequalities by simplifying and lining up the common variables. (There is no need to solve here.)

Simplify the list: $u < t$, $r < b$, $f < t$, and $t < r$.

Then, line up the variables...

$u < t$
$r < b$
$f < t$
$t < r$

...and combine.

$u < t < r < b$
$f < t$

Notice that, when LINING UP terms, it is best to focus first on the most frequently used term. In this case, t is the most frequently used term, so the inequalities are lined up around t. The remaining inequality, $r < b$, is lined up afterwards, around the r that it shares with the last inequality.

When working with variables, it is not always possible to combine all the inequalities, as we see in this example. We cannot really fit the inequality $f < t$ into the long combination. We do know that both u and f are less than t, but we do not know the relationship between u and f.

We can see from our combination that the answer to the question is YES: b is greater than u.

Plugging in the Extremes

One of the most effective ways of solving GMAT inequality problems is to focus on the EXTREME VALUES of a given inequality. This is particularly helpful when solving problems that involve multiple inequalities. For example:

> **Given that $0 \leq x \leq 3$, and $y < 8$, which of the following could NOT be the value of xy?**
>
> **(A) 0 (B) 8 (C) 12 (D) 16 (E) 24**

To solve this problem, let us consider the EXTREME VALUES of each variable.

The lowest value for x is **0**.
The highest value for x is **3**.

Test extreme values to find the right answer.

There is no lower limit to y.
The highest value for y is **less than 8**.
(Since y can't be 8, we term this upper limit "less than 8".)

What is the lowest value for xy? Plug in the lowest values for both x and y. In this problem, y has no lower limit, so there is no lower limit to xy.

What is the highest value for xy? Plug in the highest values for both x and y. In this problem, the highest value for x is **3**, and the highest value for y is **less than 8**.

Multiplying these two extremes together yields: (3) × (less than 8) = less than 24.

Thus, xy CANNOT be 24, and the answer is **(E)**.

Problem Set

1. If $4x - 12 \geq x + 9$, which of the following must be true?
   (A) $x > 6$ (B) $x < 7$ (C) $x > 7$ (D) $x > 8$ (E) $x < 8$

2. Which of the following is equivalent to $-3x + 7 \leq 2x + 32$?
   (A) $x \geq -5$ (B) $x \geq 5$ (C) $x \leq 5$ (D) $x \leq -5$

3. If $G^2 < G$ and G is positive, which of the following could be G?
   (A) 1 (B) $\frac{23}{7}$ (C) $\frac{7}{23}$ (D) -4 (E) -2

4. If $5B > 4B + 1$, is $B^2 > 1$?

5. If $x > y$, $x < 6$, and $y > -3$, what is the largest prime number that could be equal to $x + y$?

6. If $|A| > 19$, which of the following could not be equal to A?
   (A) 26 (B) 22 (C) 18 (D) -20 (E) -24

7. If $|10y - 4| > 7$ and $y < 1$, which of the following could be y?
   (A) $-.8$ (B) $-.1$ (C) $.1$ (D) 0 (E) 1

8. If $a > 7$, $a + 4 > 13$, and $2a < 30$, which of the following must be true?
   (A) $9 < a < 15$ (B) $11 < a < 15$ (C) $15 < a < 20$ (D) $13 < a < 15$

9. If $2x > 3y$ and $4y > 5z$, and z is positive, which of the following must be true?
   (A) $x < y < z$ (B) $2x < 3y < 5$ (C) $2x > 4y > 5z$ (D) $8x > 12y > 15z$

10. Given that a, b, c, and d are integers, and that $a < b$, $d > c$, $c > b$, and $d - a = 3$, which of the following must be true?
    (A) $b = 2$ (B) $d - b = 3$ (C) $c - a = 1$ (D) $c - b = 1$ (E) $a + b = c + d$

11. If $d > a$ and $L < a$, which of the following cannot be true?
    (A) $d + L = 14$ (B) $d - L = 7$ (C) $d - L = 1$ (D) $a - d = 9$ (E) $a + d = 9$

12. If $a^2b > 1$ and $b \leq 2$, which of the following could be the value of a?
    (A) $\frac{1}{2}$ (B) $\frac{1}{4}$ (C) $-\frac{1}{2}$ (D) -2 (E) $\frac{2}{3}$

13. If $TG > 0$ and G is negative, which of the following must be positive?
    (A) $T + G$ (B) $2T + G$ (C) $T - G$ (D) $G - T$ (E) $-G - T$

14. If $\frac{AB}{7} > \frac{1}{14}$ and $A = B$, which of the following must be greater than 1?

(A) $A + B$ (B) $1 - A$ (C) $2A^2$ (D) $A^2 - \frac{1}{2}$ (E) A

15. If $B^3A < 0$ and $A > 0$, which of the following must be negative?

(A) AB (B) B^2A (C) B^4 (D) $\frac{A}{B^2}$ (E) $-\frac{B}{A}$

1. **(A):** $4x - 12 \geq x + 9$
$3x \geq 21$
$x \geq 7$

If $x \geq 7$, then $x > 6$.

2. **(A):** $-3x + 7 \leq 2x + 32$
$-5x \leq 25$
$x \geq -5$

When you divide by a negative number, you must reverse the direction of the inequality symbol.

3. **(C):** Eliminate (D) and (E), since they violate the given condition that G is positive. Then test (A): 1 is not less than 1; eliminate (A). Then test (B) and (C) to see which satisfies the inequality. Alternately, you might recognize that a number decreases when multiplied by a proper fraction, so the answer must be (C).

4. **YES:** $5B > 4B + 1$
$B > 1$
The squares of all numbers greater than 1 are also greater than 1. So $B^2 > 1$.
(Don't "square both sides" of an inequality blindly. If both sides are positive, you can get away with it. If both sides are negative, you must flip the sign of the inequality. And if the sides are different or indeterminate, you can't square the inequality at all.)

5. **11:** Simplify the inequalities, so that all the inequality symbols point in the same direction. Then, line up the inequalities as shown. Finally, combine the inequalities.

$$\begin{array}{lll} y < x & & \\ \quad x < 6 & \rightarrow & -3 < y < x < 6 \\ -3 < y & & \end{array}$$

The upper extreme for x is **less than 6**. The upper extreme for y is also **less than 6**, as long as it is less than x. Therefore, $x + y$ must be **less than 12**. The largest prime number less than 12 is 11.

6. **(C):** If $|A| > 19$, then $A > 19$ OR $A < -19$. The only answer choice that does not satisfy either of these inequalities is (C) 18.

7. **(A):** First, eliminate any answer choices that do not satisfy the simpler of the two inequalities, $y < 1$. Based on this inequality alone, you can eliminate (E). Then, simplify the first inequality.

$$\begin{array}{lcl} 10y - 4 > 7 & \text{OR} & -10y + 4 > 7 \\ 10y > 11 & & 10y < -3 \\ y > 1.1 & & y < -\frac{3}{10} \end{array}$$

The only answer choice that satisfies this inequality is (A) –.8.

8. **(A):** First, solve the second and third inequalities. Simplify the inequalities, so that all the inequality symbols point in the same direction. Then, line up the inequalities as shown. Finally, combine the inequalities.

$$\begin{array}{l} 9 < a \\ \quad a < 15 \\ 7 < a \end{array} \quad \rightarrow \quad 9 < a < 15$$

Notice that all the wrong answers are more constrained: the low end is too high.

9. **(D):** First, multiply the first and second inequalities so that they have the common term $12y$.

$$4(2x > 3y) \rightarrow 8x > \mathbf{12y}$$
$$3(4y > 5z) \rightarrow \mathbf{12y} > 15z$$

Simplify the inequalities, so that all the inequality symbols point in the same direction. Then, line up the inequalities and combine.

$$\begin{array}{l} 0 < 15z \\ \quad 15z < 12y \\ \qquad 12y < 8x \end{array} \quad \rightarrow \quad 0 < 15z < 12y < 8x$$

10. **(D):** Simplify the inequalities, so that all the inequality symbols point in the same direction. Then, line up the inequalities as shown. Finally, combine the inequalities.

$$\begin{array}{l} a < b \\ \quad b < c \\ \qquad c < d \end{array} \quad \rightarrow \quad a < b < c < d$$

Since we know that $d - a = 3$, we know that the variables a, b, c, and d are consecutive integers, in ascending order. Therefore, $c - b = 1$.

11. **(D):** Simplify the inequalities, so that all the inequality symbols point in the same direction. Then, line up the inequalities as shown. Finally, combine the inequalities.

$$\begin{array}{l} L < a \\ \quad a < d \end{array} \quad \rightarrow \quad L < a < d$$

Since d is a larger number than a, $a - d$ cannot be positive. Therefore, (D) cannot be true.

12. **(D):** Plug the upper extreme value of b into the first inequality:

$$a^2(2) > 1$$
$$a^2 > \frac{1}{2}$$

(D) -2 is the only answer choice that, when squared, yields a number greater than ½.

13. **(E):** If G is negative, T must also be negative. You can generate counter-examples to eliminate each incorrect answer choice.

(A) $T + G$	$\rightarrow$	$-3 + (-4) = -7$	Incorrect
(B) $2T + G$	$\rightarrow$	$2(-1) + (-6) = -8$	Incorrect
(C) $T - G$	$\rightarrow$	$-5 - (-3) = -2$	Incorrect
(D) $G - T$	$\rightarrow$	$-7 - (-4) = -3$	Incorrect

It is impossible to generate an example in which the value of (E) is negative. $-G$ will always be a positive number, since G is negative. Subtracting a negative number (T) from a positive number ($-G$) will always yield a positive number.

14. **(C):** $\frac{AB}{7} > \frac{1}{14}$ Cross-multiply across the inequality.

$14AB > 7$ Then, divide both sides by 7.

$2AB > 1$ Since you know that $A = B$, $2AB = 2A^2$.

$2A^2 > 1$

15. **(A):** If A is positive, B^3 must be negative. Therefore, B must be negative. If A is positive and B is negative, the product AB must be negative.

Chapter 7
of
EQUATIONS, INEQUALITIES, & VIC's

VIC's

In This Chapter . . .

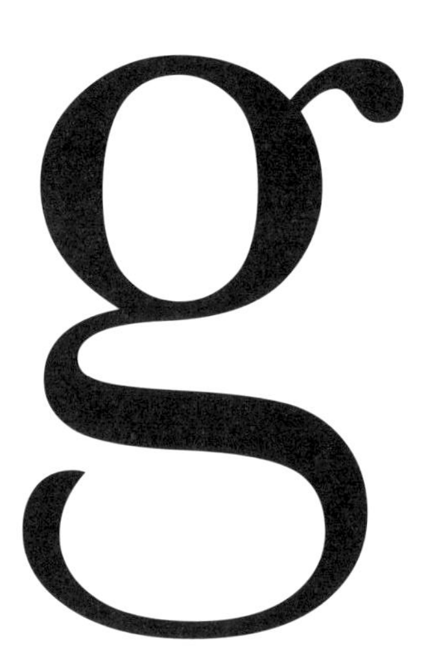

- Pick Numbers and Use a Tracking Chart
- Solve the Problem With Real Numbers
- TEACH: Test Each Answer Choice
- Equation-Based VIC's

VIC's

One final way the GMAT tests your ability to work with variables is through VIC problems. VIC problems are those that have VARIABLE EXPRESSIONS IN THE ANSWER CHOICES.

A typical VIC problem looks something like this:

Jack bought x pounds of candy at d dollars per pound. If he ate w pounds of his candy and sold the rest to Jill for m dollars per pound, how much money did Jack spend, in dollars, on the candy that he ate himself?

(A) $xd - wm$
(B) $xm - wd$
(C) $xd - xm + wm$
(D) $xd + xm - xw$
(E) $2xd - xm$

Pick numbers and use a tracking chart to solve tough VIC's.

Instead of trying to solve these seemingly complex problems algebraically, it is much easier to PICK NUMBERS for each variable using a VIC TRACKING CHART. After doing this, test each answer choice and you will find the solution.

Pick Numbers and Use a Tracking Chart

ALL VIC problems can be solved by picking numbers and using a VIC TRACKING CHART. Let us return to the Jack and Jill candy problem to see this strategy in action.

Jack bought x pounds of candy at d dollars per pound. If he ate w pounds of his candy and sold the rest to Jill for m dollars per pound, how much money did Jack spend, in dollars, on the candy that he ate himself?

(A) $xd - wm$ **(B)** $xm - wd$ **(C)** $xd - xm + wm$ **(D)** $xd + xm - xw$ **(E)** $2xd - xm$

The first step is to pick numbers for each variable and place them in a VIC Tracking Chart like the one shown below. There are several guidelines to keep in mind when picking numbers:

1. NEVER pick the numbers 1 or 0.
2. Make sure all the numbers that you pick are DIFFERENT.
3. Pick SMALL numbers.

The VIC Tracking Chart to the right shows the numbers that we arbitrarily chose for each variable. Notice that the numbers are all different, and that they are all relatively small.

Variable	Number	Description
x	10	# of pounds Jack bought
d	3	dollar price per pound in store
w	7	# of pounds Jack ate
m	2	dollar price Jill paid per pound

Solve the Problem With Real Numbers

The second step is to simply answer the question, using the numbers that we have picked.

Jack bought 10 pounds of candy at \$3 per pound. Therefore, he spent \$30.
Jack ate 7 pounds of his candy and sold the rest to Jill. Therefore, he sold 3 pounds to Jill. Jack sold the 3 pounds to Jill at \$2 per pound. Therefore, Jill paid \$6 for the candy.

How much money did Jack spend on the candy that he ate himself?

Jack spent \$30 on candy and sold \$6 worth of that candy to Jill. Therefore, Jack spent \$24 on the candy that he ate himself.

Test *each* answer choice, even if you find one that works.

Once you have arrived at a numerical answer, circle it on your scrap paper. This is your TARGET NUMBER!

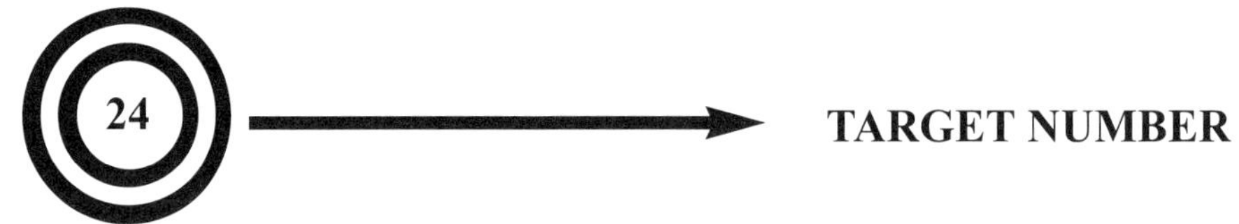

TEACH: Test Each Answer Choice

The third and final step is to test each answer choice. Plug the numbers you have picked into each answer choice. Whichever answer choice yields your target number is the correct solution to the problem.

Variable	Number	Description
x	10	# of pounds Jack bought
d	3	dollar price per pound in store
w	7	# of pounds Jack ate
m	2	dollar price Jill paid per pound

(A) $xd - wm = (10)(3) - (7)(2) = 30 - 14 = 16$ **Incorrect**
(B) $xm - wd = (10)(2) - (7)(3) = 20 - 21 = -1$ **Incorrect**
(C) $xd - xm + wm = (10)(3) - (10)(2) + (7)(2) = 30 - 20 + 14 = 24$ **CORRECT**
(D) $xd + xm - xw = (10)(3) + (10)(2) - (10)(7) = 30 + 20 - 70 = -20$ **Incorrect**
(E) $2xd - xm = (2)(10)(3) - (10)(2) = 60 - 20 = 40$ **Incorrect**

The correct answer is (C), as this is the only choice which yields the target number, 24.

When solving VIC's, keep in mind that **you must test <u>every</u> answer choice, even if you have already found one that works.**

Why? Sometimes you will find that MORE THAN ONE ANSWER CHOICE yields your target number. In such cases, you should pick new numbers and test these remaining answer choices again, until only one answer choice yields your target number.

Equation-Based VIC's

The preceding example with Jack and Jill represents the classic type of VIC problem: a situation-based word problem with numerous variables. There is, however, another type of VIC—the equation-based VIC problem. These problems give you an equation and ask you to choose an expression that is equivalent to one (or a combination) of the variables in the equation. For example:

If $\frac{abc}{72} = \frac{2}{d}$, which of the following expressions is equivalent to $ab - 2$?

(A) 72 (B) $\frac{72}{cd}$ (C) $\frac{144}{cd}$ (D) $\frac{a(144-2cd)}{cda}$ (E) $\frac{144a-2cd}{cd}$

Only pick numbers for the variables on one side of the equation. Then, solve the equation to figure out the variables on the other side.

When picking numbers for equation-based VIC's, only pick numbers for variables on one side of the equation. Then, solve the equation to figure out the remaining variables.

variable	number
a	2
b	3
c	4
d	6

We decide that a is 2, b is 3, and c is 4. Plugging these into the equation determines d.

$$\frac{abc}{72} = \frac{2}{d} \rightarrow \frac{(2)(3)(4)}{72} = \frac{2}{d} \rightarrow \frac{24}{72} = \frac{2}{d} \rightarrow \frac{1}{3} = \frac{2}{d} \rightarrow d = 6$$

Then, we find the target number by plugging the numbers we have selected into the expression $ab - 2$: $(2)(3) - 2 = 6 - 2 = 4$.

Use the target number to test each answer choice. We plug in 2 for a, 3 for b, 4 for c, and 6 for d, in search of our target number: 4.

(A) 72 **Incorrect**

(B) $\frac{72}{cd} = \frac{72}{(4)(6)} = \frac{72}{24} = 3$ **Incorrect**

(C) $\frac{144}{cd} = \frac{144}{(4)(6)} = \frac{144}{24} = 6$ **Incorrect**

(D) $\frac{a(144-2cd)}{cda} = \frac{2(144-(2)(4)(6))}{(4)(6)(2)} = \frac{(2)(96)}{48} = 4$ **CORRECT**

(E) $\frac{144a-2cd}{cd} = \frac{(144)(2)-(2)(4)(6)}{(4)(6)} = \frac{240}{24} = 10$ **Incorrect**

Just as with other VIC's, be sure to test every answer choice to make sure that only one works.

Problem Set

Solve each problem with a tracking chart.

1. If x, y, and z are consecutive integers, which of the following must be an integer?
(A) $\frac{x+2y+2z}{3}$ (B) $\frac{x+y+z}{3}$ (C) $\frac{x+y+z}{2}$ (D) $\frac{x+y+z}{6}$

2. Maria will be x years old in 12 years. How old was she 9 years ago?
(A) $x-3$ (B) $x-20$ (C) $x+3$ (D) $x-21$ (E) $x+6$

3. If Cecil reads T pages per minute, how many hours will it take him to read 500 pages?
(A) $\frac{500}{T}$ (B) $\frac{500}{60T}$ (C) $\frac{5T}{6}$ (D) $3000T$ (E) $\frac{60T}{500}$

4. A town's oldest inhabitant is x years older than the sum of the ages of the Yow triplets. If the oldest inhabitant is now J years old, how old will one of the triplets be in 20 years?
(A) $\frac{J-50}{3}$ (B) $\frac{3(J+20)}{x}$ (C) $\frac{J+x-50}{3}$ (D) $\frac{J-x+60}{3}$ (E) $\frac{J+x-20}{3}$

5. K is an even number, and G is an integer. Which of the following cannot be odd?
(A) $K+G$ (B) KG (C) $K-G$ (D) $2K+G$ (E) $3(K+G)$

6. Mr. and Mrs. Wiley have a child every J years. Their oldest child is now T years old. If they have a child 2 years from now, how many children will they have in total?
(A) $\frac{T+2}{J}+1$ (B) $JT+1$ (C) $\frac{J}{T}+\frac{1}{T}$ (D) $TJ-1$ (E) $\frac{J+T}{J}$

7. If $b=\frac{3a-6}{4}$, then what is a?
(A) $\frac{12b}{5}$ (B) $\frac{4b-6}{3}$ (C) $\frac{3b+6}{4}$ (D) $\frac{4b+6}{3}$ (E) $3b+6$

8. If $H=\frac{x^3-6x^2-x+30}{x-5}$ and $x \neq 5$, then H is equivalent to which of the following?
(A) x^2-x-6 (B) x^3+3x^2+3x (C) x^3-25 (D) x^3-5x^2-3x (E) x^2+x+10

9. If $\frac{N}{T}=\frac{P}{M}$, then MT is equivalent to which of the following?
(A) NP (B) $\frac{NP}{T}$ (C) $\frac{MN}{T}$ (D) $\frac{NP}{M}$ (E) $\frac{PT^2}{N}$

10. The average of A, B, and C is D. What is the average of A and B?
(A) $\frac{A+B-C}{3}$ (B) $\frac{3D-C}{2}$ (C) $\frac{3D-A-B+C}{3}$ (D) $\frac{3D+A+B+C}{4}$ (E) 9

11. If $a = 20bc$, then a is what percent of b?

(A) $20c$ (B) $2000c$ (C) $\frac{c}{20}$ (D) $\frac{c}{2000}$ (E) $c + 20$

12. If Tessie wins M dollars in a billiards tournament and J dollars at each of seven county fairs, what is her average prize (in dollars)?

(A) $\frac{M+7J}{8}$ (B) $\frac{M+J}{7}$ (C) $\frac{M+7J}{7}$ (D) $\frac{7M+J}{8}$ (E) $M + \frac{J}{7}$

13. If $\frac{x^2 - 11x + 28}{L} = 9$, which of the following is equivalent to L?

(A) $L(x^2 - 11x + 19)$

(B) $\frac{9}{x^2 + 11x - 28}$

(C) $\frac{(x-7)(x-4)}{9}$

(D) $\frac{(x+7)(x+4)}{9x^2 + 28}$

14. Kate can run A feet in B seconds. Amelia can run C feet in D seconds. In a race of 100 feet, by how many seconds will Kate beat Amelia (assuming they both run at these rates)?

(A) $\frac{100(AD - BC)}{AC}$ (B) $\frac{100BC - 100DA}{AC}$ (C) $\frac{AB - CD}{100}$ (D) $\frac{AD - CB}{100}$

15. X percent of Y percent of Z is decreased by Y percent. What is the result?

(A) $\frac{100XYZ - XY^2Z}{1{,}000{,}000}$

(B) $\frac{XZ - Y}{100}$

(C) $\frac{XZ - Y}{10{,}000}$

(D) $\frac{XYZ - 2Y}{100}$

(E) $\frac{XYZ - 2Y}{10{,}000}$

1. **(B):** Assign values to each variable, and record them in a tracking chart. Then, test each answer choice.

variable	number
x	2
y	3
z	4

Note that we must pick *consecutive* integers, as specified in the question.

(A) $\frac{x+2y+2z}{3} = \frac{2+2(3)+2(4)}{3} = \frac{16}{3}$

(B) $\frac{x+y+z}{3} = \frac{2+3+4}{3} = \frac{9}{3} = 3$

(C) $\frac{x+y+z}{2} = \frac{2+3+4}{2} = \frac{9}{2}$

(D) $\frac{x+y+z}{6} = \frac{2+3+4}{6} = \frac{9}{6} = \frac{3}{2}$

The only answer choice that yields an integer is (B).

2. **(D):** Let $x = 50$. If Maria will be 50 years old in 12 years, she is currently 38 years old. Therefore, she was 29 years old 9 years ago. Test each answer choice to find the one that yields the target number 29.

(A) $x - 3 = 50 - 3 = 47$ Incorrect
(B) $x - 20 = 50 - 20 = 30$ Incorrect
(C) $x + 3 = 50 + 3 = 53$ Incorrect
(D) $x - 21 = 50 - 21 = 29$ CORRECT
(E) $x + 6 = 50 + 6 = 56$ Incorrect

3. **(B):** Let $T = 10$. If Cecil reads 10 pages per minute, he will read a 500-page novel in 50 minutes. This is equivalent to 50/60, or 5/6 hours. Test each answer choice to find the one that yields the target number 5/6.

(A) $\frac{500}{T} = \frac{500}{10} = 50$ Incorrect

(B) $\frac{500}{60T} = \frac{500}{(60)(10)} = \frac{5}{6}$ CORRECT

(C) $\frac{5T}{6} = \frac{5(10)}{6} = \frac{50}{6}$ Incorrect

(D) $3000T = 3000(10) = 30{,}000$ Incorrect

(E) $\frac{60T}{500} = \frac{60(10)}{500} = \frac{6}{5}$ Incorrect

4. **(D):** If the oldest inhabitant is 100 years old, and he is 10 years older than the sum of the ages of the Yow triplets, each triplet is 30 years old. In 20 years, each triplet will be 50 years old. Test each answer choice to find the one that yields the target number 50.

variable	number
x	10
J	100

(A) $\frac{J-50}{3}=\frac{100-50}{3}=\frac{50}{3}$ Incorrect

(B) $\frac{3(J+20)}{x}=\frac{3(100+20)}{10}=36$ Incorrect

(C) $\frac{J+x-50}{3}=\frac{100+10-50}{3}=\frac{60}{3}=20$ Incorrect

(D) $\frac{J-x+60}{3}=\frac{100-10+60}{3}=50$ CORRECT

(E) $\frac{J+x-20}{3}=\frac{100+10-20}{3}=30$ Incorrect

5. **(B):** Select an even number for K and an integer for G. Then, test each answer choice to find the one that does *not* yield an odd number.

variable	number
K	4
G	3

(A) $K+G=4+3=7$ Incorrect
(B) $KG=4\times 3=12$ CORRECT
(C) $K-G=4-3=1$ Incorrect
(D) $2K+G=2(4)+3=11$ Incorrect
(E) $3(K+G)=3(4+3)=21$ Incorrect

6. **(A):** If the Wileys have a child every 2 years and the oldest child is 12, they have 7 children, ages 0 (just born), 2, 4, 6, 8, 10, and 12. If they have another child 2 years from now, they will have 8 children. Test each answer choice to find the one that yields the target number 8.

variable	number
J	2
T	12

(A) $\frac{T+2}{J}+1=\frac{12+2}{2}+1=8$ CORRECT

(B) $JT+1=24+1=25$ Incorrect

(C) $\frac{J}{T}+\frac{1}{T}=\frac{2}{12}+\frac{1}{12}=\frac{1}{4}$ Incorrect

(D) $TJ-1=24-1=23$ Incorrect

(E) $\frac{J+T}{J}=\frac{2+12}{2}=7$ Incorrect

7. **(D):** Assign a value to a first. Then, plug that value into the equation to find the value of b.

Let $a = 6$.

$$b = \frac{3(6) - 6}{4} = 3$$

variable	number
a	6
b	3

Test each answer choice to find the one that yields the correct value for a, 6.

(A) $\frac{12b}{5} = \frac{12(3)}{5} = \frac{36}{5}$ Incorrect

(B) $\frac{4b-6}{3} = \frac{4(3)-6}{3} = 2$ Incorrect

(C) $\frac{3b+6}{4} = \frac{3(3)+6}{4} = \frac{15}{4}$ Incorrect

(D) $\frac{4b+6}{3} = \frac{4(3)+6}{3} = 6$ CORRECT

(E) $3b + 6 = 3(3) + 6 = 15$ Incorrect

8. **(A):** Assign a value to x first. Then, plug that value into the equation to find the value of H.

Let $x = 2$.

$$H = \frac{2^3 - 6(2)^2 - 2 + 30}{2 - 5} = -4$$

variable	number
x	2
H	-4

Test each answer choice to find the one that yields the correct value for H, -4.

(A) $x^2 - x - 6 = 2^2 - 2 - 6 = -4$ CORRECT
(B) $x^3 + 3x^2 + 3x = 2^3 + 3(2)^2 + 3(2) = 8 + 12 + 6 = 26$ Incorrect
(C) $x^3 - 25 = 2^3 - 25 = -17$ Incorrect
(D) $x^3 - 5x^2 - 3x = 2^3 - 5(2)^2 - 3(2) = 8 - 20 - 6 = -18$ Incorrect
(E) $x^2 + x + 10 = 2^2 + 2 + 10 = 4 + 2 + 10 = 16$ Incorrect

9. **(E):** Select variables that will satisfy the equation $\frac{N}{T} = \frac{P}{M}$. If we select the values shown in the tracking chart to the right, $MT = 3 \times 2 = 6$. Then, test each answer choice to find the one that yields the target number 6.

variable	number
N	6
T	2
P	9
M	3

(A) $NP = 6 \times 9 = 54$ Incorrect

(B) $\frac{NP}{T} = \frac{6 \times 9}{2} = 27$ Incorrect

(C) $\frac{MN}{T} = \frac{3 \times 6}{2} = 9$ Incorrect

(D) $\frac{NP}{M} = \frac{6 \times 9}{3} = 18$ Incorrect

(E) $\frac{PT^2}{N} = \frac{9 \times 2^2}{6} = 6$ CORRECT

Depending on the numbers you pick, you are likely to find more than one answer choice that yields the target number. If this happens, pick different numbers and try again. The answer choice that yields the target number both times is the correct answer.

10. **(B):** If the average of A, B, and C is D, then $\frac{A+B+C}{3} = D$. If we select the numbers shown in the tracking chart, the average of A and B is 3.5. Test each answer choice to find the one that yields the target number 3.5.

variable	number
A	3
B	4
C	8
D	5

(A) $\frac{A+B-C}{3} = \frac{3+4-8}{3} = -\frac{1}{3}$ Incorrect

(B) $\frac{3D-C}{2} = \frac{3(5)-8}{2} = \frac{7}{2} = 3.5$ CORRECT

(C) $\frac{3D-A-B+C}{3} = \frac{3(5)-3-4+8}{3} = \frac{16}{3}$ Incorrect

(D) $\frac{3D+A+B+C}{4} = \frac{3(5)+3+4+8}{4} = \frac{30}{4}$ Incorrect

(E) 9 Incorrect

Avoid selecting consecutive integers to represent A, B, and C. If you do this, B and D will have the same value. Remember, you always want to pick different numbers for all the variables in the problem.

11. **(B):** Select values for a, b, and c that make the equation $a = 20bc$ true, as shown in the tracking chart to the right. Since $120 \div 2 = 60$, a is 6000% of b. Test each answer choice to find the one that yields the target number 6000.

variable	number
a	120
b	2
c	3

(A) $20c = 20 \times 3 = 60$ Incorrect

(B) $2000c = 2000 \times 3 = 6000$ CORRECT

(C) $\frac{c}{20} = \frac{3}{20}$ Incorrect

(D) $\frac{c}{2000} = \frac{3}{2000}$ Incorrect

(E) $c + 20 = 3 + 20 = 23$ Incorrect

12. **(A):** If Tessie wins 10 dollars in a billiards tournament and 20 dollars at each of seven county fairs, her average prize is $\frac{10+(7)(20)}{8} = 18.75$. You might recognize this as choice A.

variable	number
M	10
J	20

However, you can also solve the problem by testing each answer choice.

(A) $\frac{M+7J}{8} = 18.75$ CORRECT

(B) $\frac{M+J}{7} = \frac{30}{7}$ Incorrect

(C) $\frac{M+7J}{7} = \frac{150}{7}$ Incorrect

(D) $\frac{7M+J}{8} = 11.25$ Incorrect

(E) $M + \frac{J}{7} = \frac{90}{7}$ Incorrect

13. **(C)**: $\frac{x^2 - 11x + 28}{L} = 9$

$L = \frac{x^2 - 11x + 28}{9}$

variable	number
x	3
L	$\frac{4}{9}$

If $x = 3$, then $L = \frac{4}{9}$. Test each answer choice to find the one that yields the target number $\frac{4}{9}$.

(A) $L(x^2 - 11x + 19) = \frac{4}{9}(9 - 33 + 19) = -\frac{20}{9}$ Incorrect

(B) $\frac{9}{x^2 + 11x - 28} = \frac{9}{9 + 33 - 28} = \frac{9}{14}$ Incorrect

(C) $\frac{(x-7)(x-4)}{9} = \frac{(-4)(-1)}{9} = \frac{4}{9}$ CORRECT

(D) $\frac{(x+7)(x+4)}{9x^2 + 28} = \frac{(10)(7)}{81 + 28} = \frac{70}{109}$ Incorrect

14. **(A):** Assign values to variables A, B, C, and D so that Kate's rate $\left(\frac{A}{B}\right)$ is greater than Amelia's rate $\left(\frac{C}{D}\right)$.

Use the equation $d = rt$ to calculate how fast Kate and Amelia will each run 100 feet.

variable	number
A	10
B	2
C	12
D	3

Kate: $100 = 5t$
$t = 20$ s

Amelia: $100 = 4t$
$t = 25$ s

Kate will beat Amelia by 5 seconds. Test each answer choice to find the one that yields the target number of 5.

(A) $\frac{100(AD - BC)}{AC} = \frac{100(30 - 24)}{120} = 5$ CORRECT

(B) $\frac{100BC - 100DA}{AC} = \frac{2400 - 3000}{120} = -5$ Incorrect

(C) $\frac{AB - CD}{100} = \frac{20 - 36}{100} = -.16$ Incorrect

(D) $\frac{AD - CB}{100} = \frac{30 - 24}{100} = .06$ Incorrect

15. **(A):** First, assign numbers to represent X, Y, and Z. Pick numbers that translate easily into percents.

variable	number
X	10
Y	50
Z	100

Y% of Z = 50% of 100 = 50
X% of Y% of Z = 10% of 50 = 5
Decreasing this result by Y%, or by 50%, yields 2.5.

Test each answer choice to find the one that yields the target number 2.5

(A) $\frac{100XYZ - XY^2Z}{1{,}000{,}000} = \frac{100(10)(50)(100) - (10)(2500)(100)}{1{,}000{,}000} = 2.5$ CORRECT

(B) $\frac{XZ - Y}{100} = \frac{(10)(100) - 50}{100} = 9.5$ Incorrect

(C) $\frac{XZ - Y}{10{,}000} = \frac{(10)(100) - 50}{10{,}000} = .095$ Incorrect

(D) $\frac{XYZ - 2Y}{100} = \frac{(10)(50)(100) - 2(50)}{100} = 499$ Incorrect

(E) $\frac{XYZ - 2Y}{10{,}000} = \frac{(10)(50)(100) - 2(50)}{10{,}000} = 4.99$ Incorrect

Chapter 8
of
EQUATIONS, INEQUALITIES, & VIC's

STRATEGIES FOR DATA SUFFICIENCY

In This Chapter . . .

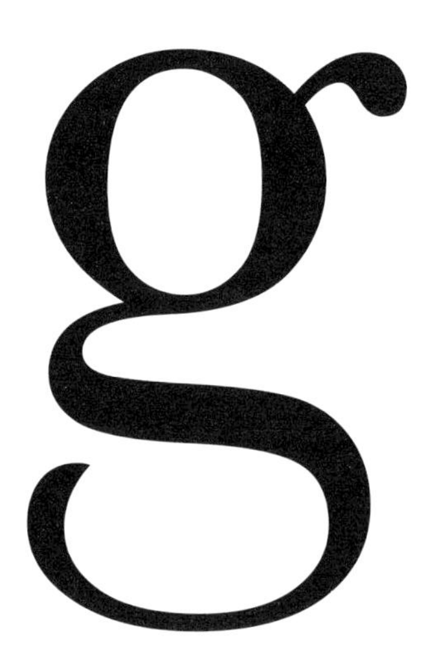

- Rephrasing: MUD Manipulations
- Sample Rephrasings for Challenging Problems

Rephrasing: MUD Manipulations

Data sufficiency problems that involve algebraic equations and inequalities can usually be solved through algebraic manipulations, such as the ones we have covered in this strategy guide. In some cases, you will need to manipulate the original question; in others, you will need to manipulate the statements. Sometimes, you will need to manipulate both.

Remember, the major manipulations include:

Multiplication and division by some number
Unsquaring and squaring
Distributing and factoring

Remember the three MUD manipulations.

You can also consider combination and substitution as manipulations for combining two or more equations. Likewise, combining inequalities is also a manipulation. When you use manipulations to rephrase, you will often uncover very simple questions that have been disguised by the GMAT writers to look complicated.

Is $p > q$?

(1) $-3p < -3q$
(2) $p - r > q - r$

A Statement (1) ALONE is sufficient, but statement (2) alone is not sufficient.
B Statement (2) ALONE is sufficient, but statement (1) alone is not sufficient.
C BOTH statements TOGETHER are sufficient, but NEITHER statement ALONE is sufficient.
D EACH statement ALONE is sufficient.
E Statements (1) and (2) together are NOT sufficient.

You can rephrase both statements by performing simple manipulations.

Rephrase statement (1) by dividing both sides by -3. The result is $p > q$, since you must switch the direction of an inequality when multiplying or dividing it by a negative number.

Rephrase statement (2) by adding r to both sides. The result is $p > q$.

Now this question is simply: **Is $p > q$?**

(1) $p > q$
(2) $p > q$

Clearly, either statement is sufficient to answer the question, since the rephrased statements exactly match the question. The answer to this data sufficiency problem is (D): EACH statement ALONE is sufficient.

Sometimes, rephrasing a statement may not uncover the answer to the question itself, but may get you closer to the information that you need to answer the question.

Consider this example:

What is the value of $r + u$?

(1) $rs - ut = 8 + rt - us$
(2) $s - t = 6$

When you see a complicated variable expression in a data sufficiency problem, look for ways to manipulate it.

As you rephrase, always keep in mind the variable or variable combo you are seeking to isolate. This can help guide how you manipulate equations and inequalities. In order to determine the value of the variable combo $r + u$, we will need to isolate $r + u$ on one side of an equation.

Manipulate statement (1) by moving all the variables to one side and factoring out common terms.

$$rs - rt + us - ut = 8$$
$$r(s - t) + u(s - t) = 8$$
$$(r + u)(s - t) = 8$$

$$r + u = \frac{8}{s - t}$$

We have manipulated statement (1) so that $r + u$ is isolated. Although we still do not have a value for $r + u$, the information uncovered by this manipulation becomes important once we look at statement (2).

On its own, statement (2) is insufficient because it tells us nothing about the value of $r + u$. It simply tells us that $s - t = 6$.

However, when we look at both statements together, we can plug the value of $s - t$ provided by statement (2) into our rephrased statement (1) to get a value for $r + u$.

$$r + u = \frac{8}{s - t} = \frac{8}{6}$$

The answer to this data sufficiency problem is (C): BOTH statements TOGETHER are sufficient, but NEITHER statement ALONE is sufficient.

Consider the following problem which involves manipulating the statements and then testing numbers.

If $ab = 8$, is a greater than b?

(1) $-3b \geq -18$
(2) $2b \geq 8$

First rephrase statement (1) by manipulating the inequality $-3b \geq -18$.

If you divide both sides of the inequality in statement (1) by -3, you get $b \leq 6$. (Don't forget to flip the direction of the inequality when dividing by a negative number.)

You can test numbers to see whether knowing that $b \leq 6$ enables us to determine whether a is greater than b.

Manipulations uncover disguised information.

If $b = 2$, then $a = 4$, in which case a IS greater than b.
However, if $b = 4$, then $a = 2$, in which case a is NOT greater than b.

Thus, statement (1) is NOT sufficient to answer the question.

Next, rephrase statement (2) by manipulating the inequality $2b \geq 8$.

If you divide both sides of the inequality in statement (2) by 2, you get $b \geq 4$. Given that b is positive, a must be positive (since we are told in the question that the product ab is positive). The smallest possible value for b is 4, which would mean that a is 2. All other possible values for b are greater than 4, and so the corresponding values for a would be less than 2.

Thus, statement (2) IS sufficient to answer the question: a is NOT greater than b.

The answer to this data sufficiency problem is (B): Statement (2) ALONE is sufficient, but statement (1) alone is not sufficient.

Rephrasing: Challenge Short Set

At the very end of this book, you will find lists of EQUATIONS, INEQUALITIES, & VIC problems that have appeared on past official GMAT exams. These lists reference problems from *The Official Guide for GMAT Review, 11th Edition* and *The Official Guide for GMAT Quantitative Review* (the questions contained therein are the property of The Graduate Management Admission Council, which is not affiliated in any way with Manhattan GMAT).

As you work through the Data Sufficiency problems listed at the end of this book, be sure to focus on *rephrasing*. If possible, try to *rephrase* each question into its simplest form *before* looking at the two statements. In order to rephrase, focus on figuring out the specific information that is absolutely necessary to answer the question. After rephrasing the question, you should also try to *rephrase* each of the two statements, if possible. Rephrase each statement by simplifying the given information into its most basic form.

In order to help you practice rephrasing, we have taken the most difficult Data Sufficiency problems on *The Official Guide* problem list (these are the problem numbers listed in the "Challenge Short Set" on page 123) and have provided you with our own sample rephrasings for each question and statement. In order to evaluate how effectively you are using the rephrasing strategy, you can compare your rephrased questions and statements to our own rephrasings that appear below. Questions and statements that are significantly rephrased appear in **bold**.

Rephrasings from *The Official Guide For GMAT Review, 11th Edition*

The questions and statements that appear below are only our *rephrasings*. The original questions and statements can be found by referencing the problem numbers below in the Data Sufficiency section of *The Official Guide for GMAT Review, 11th edition* (pages 278-290).

Note: Problem numbers preceded by "D" refer to questions in the Diagnostic Test chapter of *The Official Guide for GMAT Review, 11th edition* (pages 24-25).

D30. Let x = number of \$100 certificates sold
$20 - x$ = number of \$10 certificates sold
What is $20 - x$? OR What is x?

(1) $1650 < 100x + 10(20 - x) < 1800$
$1650 < 90x < 1600$
$16\frac{1}{9} < x < 17\frac{7}{9}$

Since x is a whole number, $x = 17$.

(2) $x > 15$

D33. Is $5^{x+2} / 25 < 1$?

For this to be less than 1, the numerator 5^{x+2} must be less than the denominator.
Is $5^{x+2} < 25$?
Is $5^{x+2} < 5^2$?
Is $x + 2 < 2$?
Is $x < 0$?

(1) $\boldsymbol{x < 0}$

(2) $x < 0$

40. What is the value of n?

(1) $n^2 + n = 6$
$n^2 + n - 6 = 0$
$(n + 3)(n - 2) = 0$ $\quad$ $\boldsymbol{n = \{-3, 2\}}$

(2) $2^{2n} = 16$
$2^{2n} = 2^4$
$2n = 4$ $\quad$ $\boldsymbol{n = 2}$

56. What is the value of xy?

(1) No meaningful rephrasing can be done here. Test numbers to see that there are multiple values for x and y that make this equation true, each of which could yield a different value for xy.

(2) No meaningful rephrasing can be done here. Test numbers to see that there are multiple values for x and y that make this equation true, each of which could yield a different value for xy.

(COMBINED) $x^2 + 1 = x + 1$

$$x^2 = x$$
$$x^2 - x = 0$$
$$x(x - 1) = 0$$
$$x = \{0, 1\};\ y = \{1, 2\};\ \mathbf{xy = \{0, 2\}}$$

89. **Does ∘ mean addition, subtraction, or multiplication?**

(1) **∘ means subtraction**

(2) **∘ means subtraction**

116. **What is $a^2 + 2ab + b^2$?**

(1) $ab = 0$

(2) $\mathbf{a^2 + b^2 = 36 + 2ab}$

(COMBINED) Insert the fact that $ab = 0$ into statement (2):
$a^2 + b^2 = 36 + 2(0)$
$a^2 + b^2 = 36$

The rephrased question (**What is $a^2 + 2ab + b^2$?**) becomes **What is $(a^2 + b^2) + 2(ab)$?**
This is $36 + 2(0) = 36$.

129. Is $b - a \geq 2(3^n - 2^n)$?

(1) $b - a = 3^{n+1} - 2^{n+1}$
$b - a = (3)3^n - (2)2^n$
$\mathbf{b - a \geq (2)3^n - (2)2^n}$

(2) No meaningful rephrasing can be done here. No information is given about a or b.

131. **Is $5^k < 1000$?**

(1) $5^{k+1} > 3000$

$5(5^k) > 3000$

$\mathbf{5^k > 600}$

(2) $5^k - 5^{k-1} = 500$

$5^k(1 - 5^{-1}) = 500$

$5^k(1 - 1/5) = 500$

$5^k(4/5) = 500$

$5^k = 500(5/4)$

$\mathbf{5^k = 625}$

133. Let m = the number of members
Let d = the dollar amount contributed by each member
$md = 60$
What is m?

(1) $d = 4$

(2) $(m - 5)(d + 2) = 60$

Substitute from the question: $d = \frac{60}{m}$

$(m - 5)(\frac{60}{m} + 2) = 60$

$m2 - 5m - 150 = 0$
$(m - 15)(m + 10) = 0$
$m = -10$ or 15, but the number of members can't be negative, so $\mathbf{m = 15}$**.**

134. **Is $n - m$ a perfect square?**

(1) $\mathbf{n - m > 15}$

(2) $n = m^2 + m$
$\mathbf{n - m = m^2}$

145. In this problem, rephrase the inequality in the question twice: once with the information provided in statement (1), and again with the information provided in statement (2).

(1) $p = r$

$$\frac{1}{r} > \frac{r}{r^2 + 2}$$

Continue rephrasing for two cases: (a) r is positive AND (b) r is negative. (The sign of r will determine whether you need to flip the inequality when you multiply or divide by the variable r.)

$$r > 0: \quad r\left(\frac{1}{r} > \frac{r}{r^2 + 2}\right)$$

$$1 > \frac{r^2}{r^2 + 2}$$

Is $r^2 < r^2 + 2$? This has an absolute answer; plug $r = 0$ to see.
Is $0 < 2$? (YES)

$$r < 0: \quad r\left(\frac{1}{r} < \frac{r}{r^2 + 2}\right)$$

$$1 < \frac{r^2}{r^2 + 2}$$

Is $r^2 > r^2 + 2$? This has an absolute answer; plug $r = 0$ to see.
Is $0 > 2$? (NO)

Statement (1) yields two different answers, depending on the sign of r, and is therefore not sufficient.

(2) ***r* is positive.**

154. Is $x < 0$?

(1) $x^3(1 - x^2) < 0$
$x^3(1 + x)(1 - x) < 0$

3 numbers multiplied together yield a negative value *either* if all 3 are negative *or* if 1 is negative and 2 are positive. In this case, it is impossible that all 3 are negative, since x^3 and $(1 - x)$ cannot both be negative. Thus 1 of the numbers must be negative while 2 must be positive.

If x is between -1 and 0, then x^3 is negative, $(1 + x)$ is positive, and $(1 - x)$ is positive.
If x is greater than 1, then x^3 is positive, $(1 + x)$ is positive, and $(1 - x)$ is negative.

Either $-1 < x < 0$ or $x > 1$.

(2) $(x^2 - 1) < 0$
$(x + 1)(x - 1) < 0$

2 numbers multiplied together yield a negative value if one is positive and the other is negative.

If x is between -1 and 1, then $(x + 1)$ is positive and $(x - 1)$ is negative.

$-1 < x < 1$

Rephrasings from *The Official Guide for GMAT Quantitative Review*

The questions and statements that appear below are only our *rephrasings*. The original questions and statements can be found by referencing the problem numbers below in the Data Sufficiency section of *The Official Guide for GMAT Quantitative Review* (pages 149-157).

85. No meaningful rephrasing can be done here. Combining inequalities by addition is the best way to solve this problem (if $a > b$ and $c > d$, then $a + c > b + d$).

115. Is $(x)^y < (y)^x$?

In this problem, rephrase the inequality in the question by plugging in the information provided in statement (1).

(1) $x = y^2$
Is $(y^2)^y < (y)^{y^2}$?
Is $y^{2y} < y^{y^2}$?
Now, drop the base (y) and compare the exponents on either side of the inequality. Since this is an inequality, we have to analyze this for two cases: when y is positive and when y is negative.

For $y > 0$: Is $2y < y^2$? This is true only if $y > 2$.

Is $y > 2$?

For $y < 0$: The question cannot be easily rephrased.

However, the $y > 0$ scenario is enough to show that statement (1) is insufficient since y might not be greater than 2.

(2) $y > 2$
No information is given about x. Test numbers to see that this information does not yield a conclusive YES or NO answer.

(COMBINED) The rephrased question from the first statement (Is $y > 2$?) is answered by the information provided in the second statement ($y > 2$). [We do not need to analyze the case where $y < 0$, since the information provided in the second statement eliminates the possibility that y is negative.]

Chapter 9
of
EQUATIONS, INEQUALITIES, & VIC's

OFFICIAL GUIDE PROBLEM SETS

In This Chapter . . .

- Equations, Inequalities, & VIC's Problem Solving List from ***The Official Guides***
- Equations, Inequalities, & VIC's Data Sufficiency List from ***The Official Guides***

Practicing with REAL GMAT Problems

Now that you have completed your study of EQUATIONS, INEQUALITIES, & VIC's it is time to test your skills on problems that have actually appeared on real GMAT exams over the past several years.

The problem sets that follow are composed of questions from two books published by the Graduate Management Admission Council® (the organization that develops the official GMAT exam):

The Official Guide for GMAT Review, 11th Edition &
The Official Guide for GMAT Quantitative Review

These two books contain quantitative questions that have appeared on past official GMAT exams. (The questions contained therein are the property of The Graduate Management Admission Council, which is not affiliated in any way with Manhattan GMAT.)

Although the questions in the Official Guides have been "retired" (they will not appear on future official GMAT exams), they are great practice questions.

In order to help you practice effectively, we have categorized every problem in The Official Guides by topic and subtopic. On the following pages, you will find two categorized lists:

(1) **Problem Solving:** Lists all Problem Solving EQUATIONS, INEQUALITIES, & VIC questions contained in *The Official Guides* and categorizes them by subtopic.

(2) **Data Sufficiency:** Lists all Data Sufficiency EQUATIONS, INEQUALITIES, & VIC questions contained in *The Official Guides* and categorizes them by subtopic.

Note: Each book in Manhattan GMAT's 8-book preparation series contains its own *Official Guide* lists that pertain to the specific topic of that particular book. If you complete all the practice problems contained on the *Official Guide* lists in the back of each of the 8 Manhattan GMAT preparation books, you will have completed every single question published in *The Official Guides*. At that point, you should be ready to take your Official GMAT exam!

Problem Solving

from *The Official Guide for GMAT Review, 11th edition* (pages 20-23 & 152-186) and *The Official Guide for GMAT Quantitative Review* (pages 62-85)

Note: Problem numbers preceded by "D" refer to questions in the Diagnostic Test chapter of *The Official Guide for GMAT Review, 11th edition* (pages 20-23).

Solve each of the following problems in a notebook, making sure to demonstrate how you arrived at each answer by showing all of your work and computations. If you get stuck on a problem, look back at the EQUATIONS, INEQUALITIES, and VIC's strategies and content contained in this guide to assist you.

CHALLENGE SHORT SET

This set contains the more difficult equations, inequalities, and VIC problems from each of the content areas.

11th edition: D16, D24, 31, 35, 38, 55, 122, 124, 130, 144, 172, 178, 190, 192, 202, 205, 220, 225, 229, 230, 232, 247
Quantitative Review: 52, 77, 83, 85, 92, 104, 106, 107, 153, 155, 173

FULL PROBLEM SET

Basic Equations

11th edition: 35, 41, 69, 102, 157, 167, 205, 211, 235
Quantitative Review: 38, 68, 107, 155, 173

Exponential Equations

11th edition: 39, 55, 58, 107, 150, 171,
Quantitative Review: 7, 72, 75, 96, 106, 153, 166

Quadratic Equations

11th edition: D16, 38, 98, 144, 155, 192, 232
Quantitative Review: 18, 55, 58, 86, 121

Formulas & Functions

11th edition: D3, 67, 99, 136, 178, 190, 202, 247
Quantitative Review: 26, 78, 91, 113, 144

Inequalities

11th edition: 46, 70, 125, 130, 146, 161, 172
Quantitative Review: 3, 83, 92

VIC's

11th edition: D24, 4, 28, 31, 85, 90, 111, 120, 122, 124, 127, 158, 164, 180, 216, 220, 225, 229, 230, 245, 246
Quantitative Review: 1, 29, 32, 42, 52, 60, 69, 77, 85, 99, 104, 111, 115, 116, 118, 124, 128, 133, 146, 171, 172

Data Sufficiency

from *The Official Guide for GMAT Review, 11th edition* (pages 24-25 & 278-290) and *The Official Guide for GMAT Quantitative Review* (pages 149-157)

Note: Problem numbers preceded by "D" refer to questions in the Diagnostic Test chapter of *The Official Guide for GMAT Review, 11th edition* (pages 24-25).

Solve each of the following problems in a notebook, making sure to demonstrate how you arrived at each answer by showing all of your work and computations. If you get stuck on a problem, look back at the EQUATIONS, INEQUALITIES, AND VIC's strategies and content contained in this guide to assist you.

Practice REPHRASING both the questions and the statements by manipulating equations and inequalities. The majority of data sufficiency problems can be rephrased; however, if you have difficulty rephrasing a problem, try testing numbers to solve it.

It is especially important that you familiarize yourself with the directions for data sufficiency problems, and that you memorize the 5 fixed answer choices that accompany all data sufficiency problems.

CHALLENGE SHORT SET

This set contains the more difficult equations, inequalities, and VIC problems from each of the content areas.

11th edition: D30, D33, 40, 56, 89, 116, 129, 131, 133, 134, 145, 154
Quantitative Review: 85, 115

FULL PROBLEM SET

Basic Equations

11th edition: D35, D37, 11, 12, 21, 26, 37, 42, 48, 52, 54, 60, 67, 70, 75, 89, 105, 127
Quantitative Review: 6, 15, 21, 23, 35, 56, 60, 77, 90, 92, 102, 103, 118

Exponential Equations

11th edition: 29, 56, 122, 129, 131, 134, 142
Quantitative Review: 9, 25, 28, 96, 105, 115

Quadratic Equations

11th edition: 30, 40, 57, 97, 116, 123, 133, 137
Quantitative Review: 37, 46, 61, 79, 80

Formulas & Functions

Quantitative Review: 68

Inequalities

11th edition: D30, D33, D38, 9, 18, 20, 24, 43, 50, 64, 80, 99, 109, 112, 119, 121, 128, 139, 143, 145, 151, 154
Quantitative Review: 32, 40, 42, 43, 51, 55, 66, 67, 85, 114